The Structure and Function of
FRESH-WATER MICROBIAL COMMUNITIES

Edited by John Cairns, Jr.

Research Division Monograph 3
Virginia Polytechnic Institute and State University
Blacksburg, Virginia 24061

Foreword

The American Microscopical Society has encouraged communication between scholars through meetings which have been held nearly every year since incorporation of the Society in 1891. The meetings have been of various types and on cogent subjects, often expressing the interests of the program chairman. Dr. John Cairns, Jr., as Vice-President and Program Chairman for the 1969 meeting of the Society, in conjunction with the American Institute of Biological Sciences, organized a symposium and arranged the printing of the proceedings which comprises this book.

The 1969 program on "The Structure and Function of Fresh-Water Microbial Communities" is especially timely with our national concern for water quality and was developed with an awareness of research in progress which is directed toward an understanding of the relationships among living things in our aquatic environment. The presentations included considerable new and unpublished information which, when coupled with a review of the subject, provided more impetus to publishing the proceedings.

This book is a compilation of manuscripts prepared following the Symposium and submitted to the American Microscopical Society for review and transmittal to Dr. John Cairns, Jr., who served as Editor.

Acknowledgment is given to the University of Vermont, an institution serving the academic community in many ways; in this instance, the use of facilities for the symposium. The Virginia Polytechnic Institute and State University is especially recognized for its assistance in publishing the symposium under the guidance of Mr. J. C. Malone, Director of Research Publications. Special acknowledgment is again given to Dr. John Cairns, Jr., who as Vice-President and Program Chairman of the American Microscopical Society, implemented both the symposium and the publication.

David T. Clark
Secretary, American Microscopical Society

CONTRIBUTORS

Harold L. Allen, W. K. Kellogg Biological Station and Department of Botany and Plant Pathology, Michigan State University.

Robert Benoit, Roger Hatcher and William Green, Virginia Polytechnic Institute and State University.

John Cairns, Jr., Department of Biology, Virginia Polytechnic Institute and State University.

G. Dennis Cooke, Institute of Limnology, Department of Biological Sciences, Kent State University.

David G. Frey, Department of Zoology, Indiana University.

John E. Hobbie, Department of Zoology, North Carolina State University.

Bassett Maguire, Jr., University of Texas at Austin.

Ross E. McKinney, Department of Civil Engineering, University of Kansas.

Howard T. Odum, Scott W. Nixon and Louis H. Disalvo, University of North Carolina.

Bruce C. Parker and Mary Ann Wachtel, Department of Botany, Washington University, St. Louis, Missouri.

Robert A. Paterson, Department of Biology, Virginia Polytechnic Institute and State University.

Ruth Patrick, Academy of Natural Sciences of Philadelphia.

George W. Salt, Department of Zoology, University of California at Davis.

George W. Saunders, Department of Zoology, University of Michigan.

J. K. G. Silvey and J. T. Wyatt, Department of Biological Sciences, North Texas State University.

Frieda B. Taub, College of Fisheries, University of Washington.

Contents

Contents—continued

Howard T. Odum, Scott W. Nixon and Louis H. DiSalvo

Adaptations for Photoregenerative Cycling

Abstract

Ecological systems requiring productive photosynthesis and regenerative respiration to maintain smooth flowing mineral cycles may develop under circumstances in which the regenerative half of the cycle tends to be limiting.

In stable less-stressed systems such as coral reefs, complex animals and microbial communities accomplish effective mineral recycling through their specializations and diversity. Even the regeneration of limestone may be managed by microbial sediment communities located within the reef formations and under the adaptive management of burrowing animals. There are some conceptual similarities between the reef renewal and urban renewal of our cities.

However, for stressed and other systems in which organic matter and structure accumulate, recycling may be boosted by adaptations that inject special energy into catabolism and disordering, thus reducing bottlenecks in the cycles of regeneration. Aquatic systems of this type develop in stressed conditions such as brines, temperature extremes, and pollution that interfere with complex animal and microbial work. These systems may boost cycling by diverting solar energy to help break down structure and storage through photorespiration, drying with fire, and melting of permafrost. Shift from successional state to climax state may often require the addition of energies to the regenerative half of the systems and special adaptation for these processes. The general prevalence of photoregeneration along with photosynthesis in daylight cancels many environmental exchanges and interferes with efforts to measure either process by gas exchange techniques. Net gains develop only to the extent that the two processes are momentarily lagging or out of phase. There is oxygen consumption and carbon dioxide utilization at sunrise but since the processes giving rise to each are coupled they are difficult to measure independently. Compartmental systems analysis allows separation of the two processes and the simulation of electrical models provides some reversed transients of the same class as those observed in gas exchanges of oxygen and carbon dioxide.

Carotenoid pigments may be important in photoregenerative functions. If so, the ratio of yellow to green plant pigments may measure the relative use of solar energy in the two processes. When photosynthesis is limiting, chlorophyll increases; when respiration is potentially limiting, other pigments may increase to accelerate regeneration. The pigment ratio may indicate the relative switching of energy to the two half systems of (P) or (R). The pigment ratio increases as climax approaches only when the initial succession started with a regime of net production and initial conditions of high nutrients and low organic stocks. Some photorespiratory capability may characterize all ecosystems. The photorespiratory use of solar energy by brine ecosystems contrasts with the coral reef's channeling of more organic matter into regenerative food chains, both providing viable formulae for their environmental circumstance.

Introduction

Many ecological systems resemble balanced aquaria with a photosynthetic, productive process (P) balanced by a regenerative, respiratory process (R) between which many minerals and work services cycle. Even those systems that have important import and export flows have P-R modules in varying quantity. The behavior of systems of this class has been much studied with modeling and with measurements. The essence is in two storage units and one multiplicative feedback loop. Figs. 1 and 2 summarize energy flows, mineral cycles and the essence of kinetics as represented in the electrical models (Odum, Beyers and Armstrong, 1963; Milsum, 1966). Fig. 3 is a legend for the energy network symbols, each one of which represents a cluster of associated functions and differential equations.

To be well adapted to available conditions and energies is to process energies competitively with effective cycling, without blockages, accumulations, or shortages. This commentary concerns the special adaptations for effective cycling that develop under different environmental conditions. For many systems the model in Fig. 1 is much too simple to predict diurnal patterns of exchange.

Discussion

Regeneration by Microbe-Animal Cities

Many ecosystems such as coral reefs contain intricate regenerative systems of animals and microbes which the animals in part manage. With well known great diversities, high degrees of behavioral programming and much specialization by both animals and microbes, the regeneration is clean, with little organic matter accumulating or remaining uncycled. In many coral reefs in the Pacific skeletons are formed and dispersed so as to maintain a reef platform and structure at a steady state near the extreme low water spring tide level. Documentation of the great diversity of

A. Energy

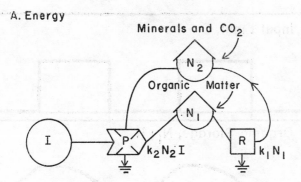

B. Mineral Cycle

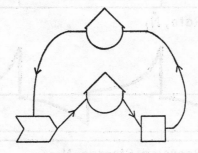

C. Flow of equations in block diagram

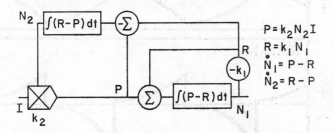

$$P = k_2 N_2 I$$
$$R = k_1 N_1$$
$$\dot{N_1} = P - R$$
$$\dot{N_2} = R - P$$

D. Passive analog

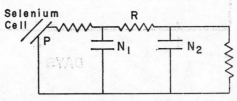

FIGURE 1. Characteristic basic system with photosynthesis and regenerative respiration balanced with respect to matter. A, Energy; B, Mineral cycle; C, Block diagram of integral equation; D, Passive electrical analog.

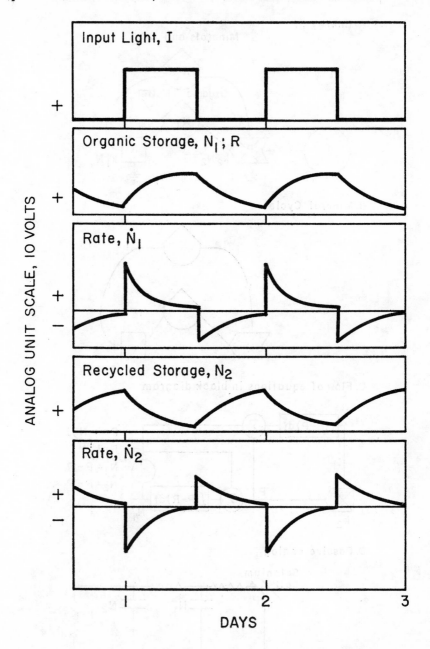

FIGURE 2. Transient responses of the basic P-R system to on and off light periods with an analog computer circuit wired to represent the block diagram of Fig. 1C, using coefficients for a terrestrial microcosm (Odum, Lugo, and Burns, 1969).

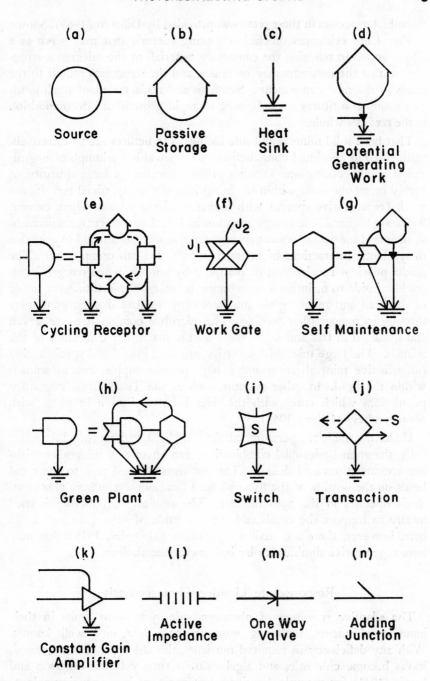

FIGURE 3. Energy network symbols from Odum and Pigeon (1969).

microbial processes in these reefs was provided by DiSalvo (1969). Shown in Fig. 4 are evidences of chitin-digesting bacteria that may serve as a zipper action in releasing the cementing materials of the calcareous structure so that the nutrients may be reused and the structures rebuilt to the needs of the new components. Synthesis and replacement of small heads as a group constitutes a steady state of rapid growth and decomposition in the reef as a whole.

That both solid mineral and soft biological structures are so effectively maintained in continual construction and renewal is a triumph of organization and specialization. DiSalvo (1969) described a large quantity of highly metabolic ooze within the interstices of the superficial reef framework (regenerative spaces) with higher respiratory rates where current levels are higher. The energy diagram in Fig. 5 is compartmentalized to show one of the overall features in regenerative metabolism and oxygen due to symbiotic interactions of infaunal animals that can pump water. The model provides acceleration of pumping by animals when oxygen levels are low inside to maintain a homeostatic balance by re-aerating the inside of the head and reducing the numbers of particulate oxygen consumers through increment filter feeding. The microbial communities are served and managed in this and other ways by the macroscopic motions of the animals. The large microbial diversity provides nutritional specialization for effective mineralization and a high protein supplement to animals within the head. In other systems, such as the Texas Bays, migratory populations which enter with the tides keep respiration in phase with photosynthesis (Odum, 1967).

Under the stable temperature, salinity, and light regimes found on many reefs, the animal-microbial combinations can effectively mineralize without accumulations and deficits. The nutrients released pass to the coral heads on the outside of the live and dead head clusters, where photosynthesis operates in the zooxanthellae. The resulting production is used locally to support the corals and the multitude of other animals. Even here, however, there are auxiliary pigments (Margalef, 1959) that may have regenerative significance by influencing metabolism.

Responses to Limiting Photosynthesis

The adaptive responses of photosynthetic systems to limits in their input requirements, including nutrients and light, are well known. With any deficiency in required nutrients, the chlorophyll is decreased, leaves become chlorotic, and algal cultures turn yellow. Emerson and Lewis (1943) for example described deficiencies of iron reducing chlorophyll, while the ratio of chlorophyll to photosynthesis was maintained.

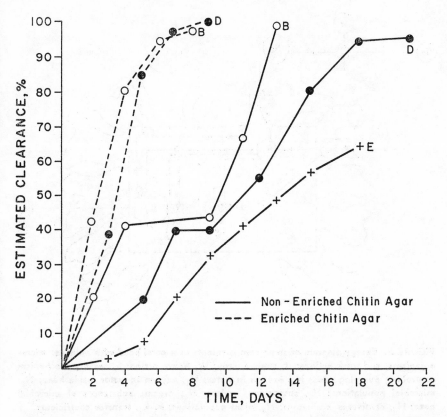

FIGURE 4. Clearance of particulate chitin in agar by coral reef microbial regenerators obtained from within skeletal spaces, providing evidence for the existence of bacteria capable of breaking down the organic (chitinous) matrix of corals; Kaneohe Bay, Hawaii. B, outer bay; D. midbay; E, inner bay.

With limitations in light, chlorophyll is increased as often described for shade adaptation. Wassink, Richardson, and Preters (1956), for example, reported shade adaptation in a deciduous tree, Gessner (1937) showed shade adaptation in land and water plants and Steeman-Nielsen (1962) showed shade adaptation in plankton.

The relationship of varying input power to constant biochemical load of downstream work processes is shown in Fig. 6. Here the Michaelis-Menton kinetics of a cyclic process interacting with an input, in this case light, provides a varying load ratio of input to output. For maximum power transfer each step involving an energy storage against a potential generating force must have a 50% energy transfer with 50% passing into dispersed heat generating entropy in the environment (Odum and Pinkerton, 1955). The system can help to regulate its input-output loading in

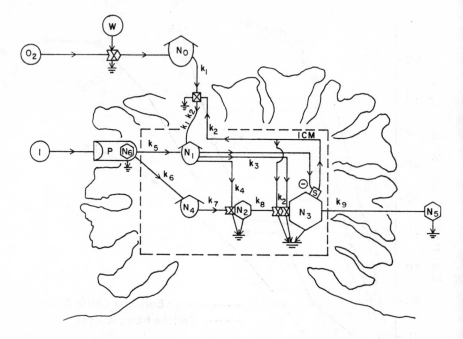

FIGURE 5. Energy diagram of some compartments of a coral head with animals, micro-organisms, and work actions: I, sun's energy; W, Wave action; S, sensory mechanism controlling pumping rates; N_0, oxygen in water; N_1, oxygen in water inside head; N_2, microbial populations; N_3, suspension feeders; N_4, organic substances of microbial ooze; N_5, carnivores, omnivores; N_6, plant populations; k_1-k_9, transfer coefficients.

the maximum power range (50% efficiency for each energy storage step) in bright light regimes by reducing its quantity of cycling receptor material, in this case chlorophyll, to that required to absorb the needed input energy load. This allows excess to pass to the next receiver units (next layer of photosynthetic tissue). Each subsequent chlorophyll-bearing unit in turn may adjust its input light absorbed to equal its output load or adjust its load. The consequence of these component mechanisms is overall use of more light at optimum efficiency for maximum power by the ecological system. The predicted theoretical relation of optimum pigment to light input for constant output is given in Fig. 7 along with an experimental example of response of *Chlorella* given by Phillips and Myers (1954). These graphs show the shade adaptation increase of chlorophyll, which tends to maintain constant output load.

The adaptive chlorophyll system serving as a gear increases with decreased input or increased output. In rich nutritive solution specialists

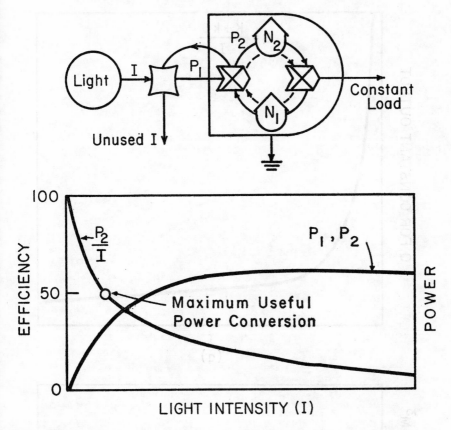

FIGURE 6. Efficiency of Michaelis-Menton cyclic modules under constant output load and varying input.

at rapid growth are not limited downstream through mechanisms such as nutrient regeneration and may operate with high chlorophyll. If, however, there is very bright light, they may be relatively limited by the high input-output ratio and reduce chlorophyll. The chlorophyll seems to be servo-regulated to adjust the load ratio. When a system has inadequate light it turns chlorophyll up, but if its nutrients are limiting, the shortage may be solved by fertilization from outside or by some more effective method of recycling, the energy for which may also have to come from outside.

One of the adaptations of specialized algae such as deep-water reds to low light intensities is a provision for a second pigment system which contributes to photosynthesis. French and Fork (1961) studied the metabolic transients that result from the interplay of two-pigment sys-

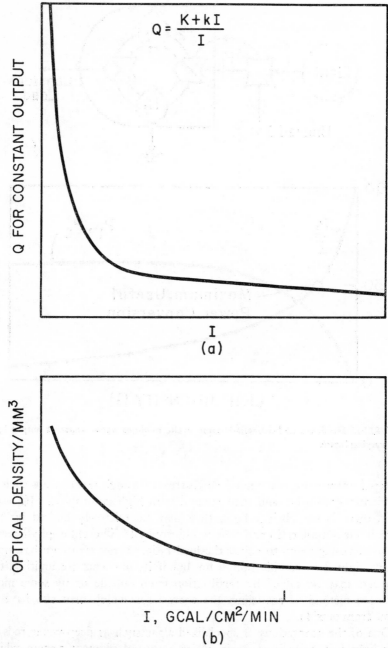

$$Q = \frac{K + kI}{I}$$

Q FOR CONSTANT OUTPUT

I

(a)

OPTICAL DENSITY/MM^3

I, GCAL/CM^2/MIN

(b)

FIGURE 7. Chlorophyll adaptations to light for maintaining steady output load.
a. Theoretical from manipulation of Michaelis-Menton equation;
b. Observed by Phillips and Myers (1954) for *Chlorella*.

tems, developing analog circuits to relate models to transient consequences which were compared with observed data. In a similar way, the arguments which follow on photoregeneration concern analog analysis of transients involving the balance of two-pigment systems, except that one-pigment is suggested as photoregenerative.

Problems of Systems Limited by
Slow Respiration

Unlike the coral reef example, many ecological systems have inadequate associations of microbes and animals to fully recycle their organic storages. Various kinds of stresses, such as high, low, and variable temperature; high, low and variable salinity; poisons; water shortages; and high organic input, make the adaptations of complex animal-microbial organizational patterns relatively expensive in energy costs. Many stress conditions thus lead to reduced diversities of animals and micro-organisms. In these environments organic substances accumulate. High organic levels are a part of brines, arctic communities, and toxic situations in pollution. See, for example, the hyperbolic graph of decreasing organic matter with increasing species diversity (Odum, 1967). What mechanisms tend to develop for mineral cycling when the energies of the stored matter are not adequate to operate an effective mineral recycling system?

Theory of Photoregeneration

A solution to inadequate recycling is to inject special energies into the regeneration process. If the cycle of P and R is blocked at the R process, it could be economical to divert light energy from photosynthesis to help respiration in an amount required to balance the two processes. Photoregeneration is shown in its simplest form in Fig. 8. The basic P-R model of Fig. 1 has been modified by adding one photoregeneration pathway, a multiplicative work gate. Such a mechanism requires some energy absorbing material that is coupled to the regenerative process. Also, since decomposition is in the direction of disorder, less energy may be required than in the initial photosynthesis. When we look at various ecosystems that have organic accumulations, we find pigments and other light-receiving operations which are serving as photoregenerators or may be proposed to have this role pending more experimental testing. In Table 1 are listed some systems that seem to fit the hypothesis.

Inverted Transients

In Fig. 9 from Nixon (1969) are diurnal curves of dissolved oxygen in a brine ecosystem. It was discovered that the normal pattern of day and

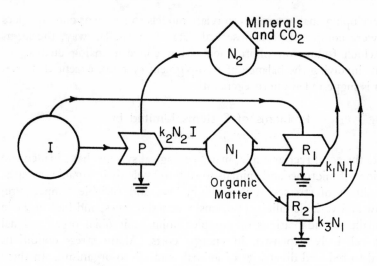

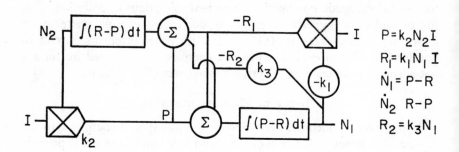

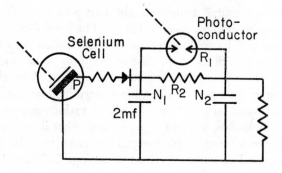

FIGURE 8. Energy diagram of basic photosynthesis and respiration with photorespiration pathway added. a. Energy diagram; b. Flow of equations in block diagram. c. Passive analog circuit.

TABLE 1.—Some Systems Which May Contain Photoregeneration.

System	Reference in which data suggest photoregeneration
Mutant *Chlorella* culture lacking chlorophyll	Kowallik (1967)
Brines	Nixon (1969)
Waste stabilization lagoons	Eppley and Macias R (1963)
Shallow marine bay bottom	Bunt (1969)
Chromogenic bacteria in lakes	Henrici (1939), Odum (1957)
Blue-green algae	Odum and Hoskin (1957), Sollins (1969)
Snow Algae	Fogg (1967)
Old *Chlorella* cultures	Fogg (1965)
Old blue-green algal cultures	Kingsbury (1956), Manny (1969)
Phytoplankton	Margalef (1959)
Desert vegetation with succulents	Ting and Dugger (1968)

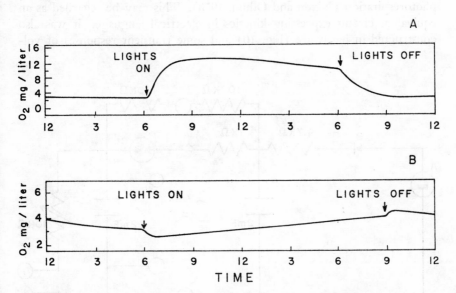

FIGURE 9. (A) Continuous diurnal curve of dissolved oxygen in a 6% salt microcosm with a healthy blue-green algal mat and moderate nutrient levels. Patterns were normal with a square wave light input from 6 a.m. to 6 p.m. (B) Continuous diurnal curve of dissolved oxygen in a 13% salt microcosm with a yellowed blue-green algal mat and very low nutrient levels. Abnormal patterns were apparent with a square wave light input from 6 a.m. to 9 p.m.

night oxygen production and consumption was missing. Often there were inverted periods, with the oxygen decreasing after lights were on and increasing briefly when lights were off. The pattern was like that of a cactus (Ting and Dugger, 1968). The pH-CO_2 was also reversed on many days. Such curves have been found in aquatic systems before, including some from the list in Table 1. The inverted pattern suggested photorespiration, a process well known in physiological work, but not appreciated for its adaptive significance to mineral cycling and ecosystem adaptation. The shape of the graphs with time suggests time lags in ordinary RC (resistor-capacitor) passive electrical circuits. It was apparent that the behavior might be accounted for by fairly simple time lags involving two storages, representing organic and inorganic pools, and a photoconductor pathway, as well as by more elaborate models with more stages of the real regeneration and production represented separately.

Passive Model

In Fig. 8 is a two compartment model with a photoconductive element representing the mechanism for the reception of the light stimulus to photorespiration (Nixon and Odum, 1970). This may be regarded as an equivalent circuit expressing kinetics in electrical language. It was also constructed in hardware (Fig. 10) and some transient responses of vol-

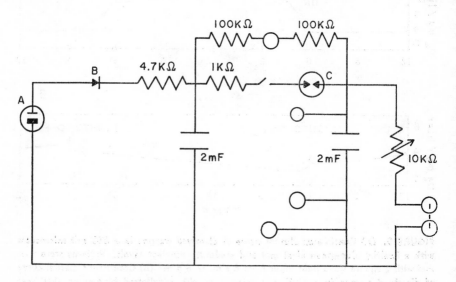

FIGURE 10. Passive electrical analog model with two storage compartments. A. Silicon photocell; B. Diode; C. Cadmium sulfide photoconductor.

tage were recorded following on and off light input. Time constants were selected by varying the resistances. As shown in Fig. 11, a transient lag was observed with some resistance settings. In the manner of the theory, changes accumulating upstream from a high resisting part of the circuit were flushed into downstream compartments by the photoconductor when light was turned on. Conceptually therefore the observed form of oxygen and carbon-dioxide release was compatible with the theory of photoregeneration as an adaptive recycling mechanism.

Operational Analog Model

Another procedure for analog simulation involves the standard operational amplifiers, which were used to model a similar network. In Fig. 12-13 a somewhat more complex theory of photoregeneration is modeled. First an energy diagram is drawn (Nixon, 1969) that is of the same class of network as the model in Fig. 10, but derives suggestion from plant physiological work on "photoheterotrophy" such as that by Bunt (1969). The energy diagram automatically defines differential equations that go

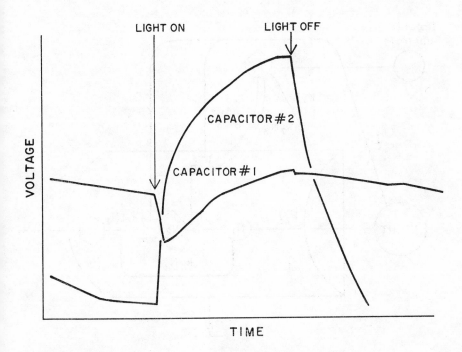

FIGURE 11. Transients from passive analog (Fig. 10) which resemble observed oxygen transients (Fig. 9).

with each module as also shown in Fig. 12. Each of these equations is block diagrammed and for each block an electrical analog component is wired (Fig. 13). Some experimental manipulation of coefficients (conductivities per unit storage equal to reciprocal of the time constant) provided the lagging transients like that in some observed gas exchange curves. An example is given in Fig. 14.

Characteristics of the Photoregenerative Model

The exercises is modeling and simulation show enough features of the observed systems to help in the interpretation of the more complex real systems which are of this class of phenomena. Where both P and R are increased by the light, only the difference in effect on outside nutrient and gas reservoirs is observed. Metabolic quantities are observed to vary with short transients affected by the history of previous storages. When photoregeneration is high, estimates of photosynthetic exchange are almost meaningless as a measure of the underlying real processes. Daytime cycles are very rapid, turning off at night with the light. Stressed systems, which have been thought to be low in primary productivity, may ac-

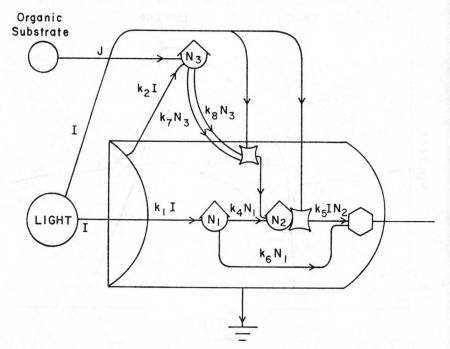

FIGURE 12. Energy diagram of photoheterotrophy from physiological suggestions (Bunt, 1969).

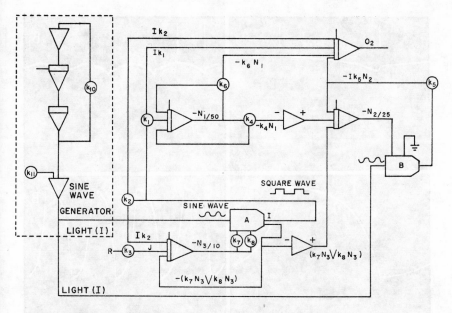

FIGURE 13. Operational analog model for energy diagram of Fig. 10.

tually be as productive as other more easily measured types. To regard cactus deserts, tropical seas, brines, etc. as low in basic productivity seems erroneous, although we don't know how erroneous until the two processes of P and R can be separated and a realistic measure of gross production obtained. The simulation procedure is one way. Direct measurement of primary photon reception under field conditions is another. Measurements with enzyme inhibitors is another explored by Nixon (1969).

Physiological Mechanisms, Carotenoid Pigments

Kowallik (1967), studying a chlorophyll-free mutant of *Chlorella*, showed data suggesting carotenoids as the receptors of the light energy for photorespiration with a hyperbolic limiting factor curve of output with intensity of blue light. This curve is consistent with the multiplier relationship in our model (Figs. 8 and 12). He cites the earlier discovery of similar phenomena by Emerson and Lewis (1943). The saturating intensity for photorespiration with *Chlorella* in blue light is one-tenth that for photosynthesis, as might be expected for a process working toward decomposition and regeneration. Light was also effective under anaerobic conditions in accelerating fermentation.

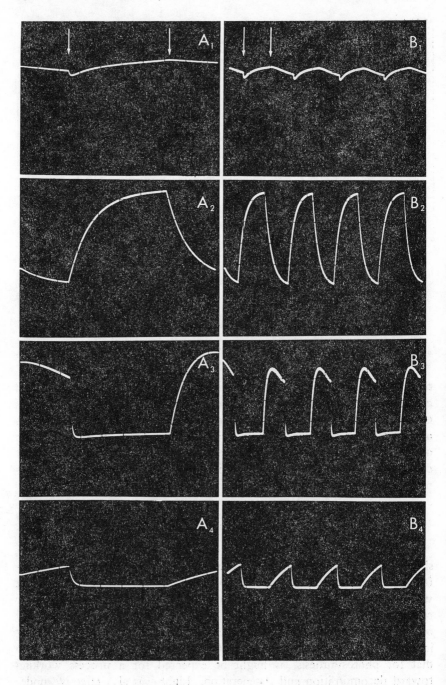

Carotenoid-Chlorophyll Ratio

The system maintaining chlorophyll has long been known to increase it when needs for photosynthesis are optimal and to decrease it when some other input is limiting. In short the chlorophyll system serves as an input load adjuster to the output chain, which is possibly consistent with other flows. When chlorophyll declines, carotenoids either increase or by comparison are left in relative predominance. There is a yellowing of plant systems when conditions for rapid net photosynthesis disappear. Margalef (1965) proposed the ratio of absorption of pigment extracts at 430 millimicrons to absorption at 665 millimicrons as a measure of ecosystem maturity in the sense of succession as well as evolutionary achievement of possible adaptations. He regarded the yellow carotenoids and other pigments as a symptom of increasing complexity of pigments and diversity of maintenance. Kingsbury (1956) and Manny (1969) correlate the ratio with nitrogen shortage.

The photoregeneration theory suggests a somewhat different interpretation. If chlorophyll is adjusting and activating when needed to keep input light load adapted to photosynthetic work chains downstream, perhaps in the same way carotenoids and other pigments of ecosystems (plants, animals, and micro-organisms) are adjusting to keep the photoregenerative energies adjusted to maintain respiration and regenerative rates. By this theory, as organic stocks accumulate with danger of becoming blockages, photoregeneration is increased either within plant species that have the mechanisms, by substitution of species, or in other microbial specialists that serve the need. The pigment ratio is thus proposed as a measure of the relative accumulation of organic and inorganic pools and of the switch-controlling diversion of light energy to the one pool which is overstocked, limiting flow downstream.

When systems are high in organic matter and photorespiration is high, the pigment ratio increases; when systems are high in inorganic materials,

---

FIGURE 14. Transients observed with active model of photorespiration.
Series A shows the values for 1 day, while series B extends the scan over 4 days to show the steady-state behavior of the model. Arrows indicate when the light goes on (left) and off (right).

A_1 and B_1 = dissolved oxygen
A_2 and B_2 = stored organics (N_1)
A_3 and B_3 = stored organics (N_2)
A_4 and B_4 = dissolved organics (N_3)

The drop in N_2 during the dark is an artifact resulting from inaccuracy in the comparator at values near zero.

the ratio goes down. By this view an increasing Margalef ratio is a measure of succession and climax only for the common case where the system starts with low stocks of organics and high concentrations of inorganic nutrients. Species may specialize for this situation simply by omission of photorespiratory systems, and it is these species that are preadapted for our fast net production demands such as algal culture, agriculture, and forestry. Zelitch (1967), for example, found photorespiration small in corn. When a steady state develops with high inorganic input and high organic output a low ratio is climax.

When systems start the other way, with high organic stocks and low inorganic stocks, the ratio is predicted to be lower at climax. When input of a steady state is high in organics and low in inorganics the ratio will be high at climax. We were able to switch a flowing stream microcosm containing a blue green algal mat (Odum and Hoskin, 1957) from green to yellow or back at will by control of the inorganic nutrition as shown in Fig. 15. In one series protozoan and flatworm populations developed, making a more effective mineral cycling system. In this series, the color remained green and the orange did not develop. Some pigment ratios are given in Table 2.

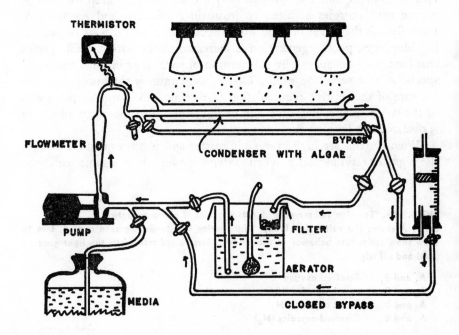

THERMISTOR

FLOWMETER

CONDENSER WITH ALGAE

BYPASS

PUMP

FILTER

AERATOR

MEDIA

CLOSED BYPASS

A

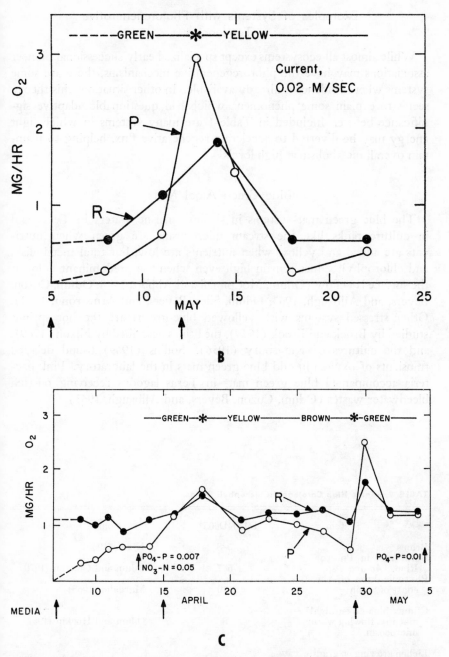

FIGURE 15. Response of blue-green algal mat color and metabolism to nutrient addition (vertical arrows) in an artificial stream microcosm (Odum and Hoskin, 1957). A: Circulating stream microcosm; B: Low current sequence; C: High current sequence.

Examples of Systems with Photoregenerative Tendencies

While almost all ecosystems except specialized early successional pioneer associations may have the photoregenerative mechanisms, there are some systems where evidence is already available. In other situations this theory seems to explain some phenomena which had questionable adaptive significance before. Included in Table 1 are many systems in which light energy may be diverted to accelerate regenerative flux, helping to maintain overall metabolism at high levels.

Blue Green Algal Mats

The blue green mat systems in shallow marine waters in Texas and in culture tanks like the stream microcosms are green when nutrients are added and yellow when nutrients are low, but total metabolism and chlorophyll values remain high even when the predominant color at the surface is yellow (Odum, McConnell and Abbott, 1958; Odum, Cuzon, Beyers, and Allbaugh, 1963; Odum, Siler, Beyers, and Armstrong, 1963). Other stressed systems with yellowed blue-greens are the hot springs studied by Brock and Brock (1969), the brines described by Nixon (1969), and the cultures of Kingsbury (1956). Sollins (1969) found delayed transients of oxygen in odd blue green mats in the laboratory. Pink bacteria accompanied blue green mats in Texas lagoons receiving oil-rich bleedwater wastes (Odum, Cuzon, Beyers, and Allbaugh, 1963).

TABLE 2.—Some High Carotenoid:Chlorophyll Ratios.

System	D430/D665*	Source
Brines	5–8	Nixon (1969)
Coral reef in Puerto Rico. *Acropora* sp.	5.7	Odum and Cintron (1970)
Phytoplankton off the coast of Spain	2–10	Margalef (1968)
Climax blue-green algal mat in a flowing water microcosm	8	Odum and Hoskin (1957)
Lichen growing on granite rock. *Cladonia* sp.	10.8	Lugo (1969)

*Optical densities of acetone extracts

Chlamydomonas in Anaerobic Waste Stabilization Lagoons

A well documented case of photoassimilation is given by Eppley and Macias (1963) where a species of *Chlamydomonas* in waste ponds in the Mohave desert used pigments to absorb and photorespire acetate produced by anaerobic bacteria. Many kinds of high organic pollutions have colored and specialized algae which may turn out to operate similarly.

Many algae such as *Chlorella* and *Haematococcus* in old or nutrient-limited cultures develop yellow pigments along with excesses of fats and other organics and their cells may increase in size. Fogg (1965) summarized these characteristics and used a colored plate to dramatize the deep green versus red phases of growing and stationary populations. These characteristics make sense mechanistically and adaptively as a switch in regimes for recycling under conditions when there is not a complex ecological community to perform the service of renewal.

A Diatom in Organic Marine Sediments

Bunt (1969) isolated and studied a diatom from organic sediments of Biscayne Bay, which was found to assimilate a variety of organic compounds at different rates in the light and in the dark. Metabolism of the assimilated organics, however, proceeded only in the light, a process Bunt termed "photoheterotrophy".

Corals

DiSalvo (1969) presented spectrophotometric curves of coral pigments from Kaneohe Bay, Hawaii. Inner bay stations stressed with terrestrial runoff and sewage pollution had higher carotenoid to chlorophyll ratios (3.4) compared to outer bay stations (2.4) and mid-bay stations (2.4). Corals with their symbiotic zooxanthellae do have abundant carotenoid pigments and adequately recycle their own photosynthetic production.

Brines

Medium and high brines have colored pigments in all their populations, including *Artemia*, halophilic bacteria, the alga *Dunaliella salina*, and the black pigmented ciliate *Fabrea* (Nixon, 1969). High brines are high in dissolved organic matter, diversity of microbes is small, and photorespiratory energy may be essential to cycling. Brines are green with blue-green algae and *D. viridis* at an earlier lower-salinity, high-nutrient stage. Nixon (1969) provides systems analyses of these systems.

Chromogenic Bacteria in Normal Waters

In the surface layers of normal types of waters, general plating of bacteria in low nutrient agar in the dark or light yields chromogens, brightly colored bacterial colonies of many kinds, which develop slowly on the agar. These are much more numerous from lighted surface waters than from dark environments. Henrici (1939) described the pattern in lakes of Wisconsin; Odum (1957) described this pattern in Silver Springs, and others have found this in the sea. DiSalvo found more pigmented bacteria near the shore in Hawaii, where organic matter was more concentrated than in Eniwetok. Are these pigmented non-photosynthetic forms pigmented for photo-regenerative purposes? If so, then photorespiration may be a normal part of most aquatic ecosystems. Many workers have mentioned occasional unpublished experiences with dark-light bottle work in rich organic waters in which light bottles have consumed more oxygen than the dark controls. The extent to which this is prevalent is the extent of an error in measuring photosynthesis by known methods.

Sitka Spruce and Permafrost

Although not photorespiration, an adaptation of dwarf Sitka Spruce in Fairbanks, Alaska may be another case of photoregeneration. There the spruce boughs are short, not spreading widely as they do farther south. The ground between trees is left exposed to summer light. If the limiting property is regeneration of minerals hindered by soil ice, the ground heating provides a shunt of light to regenerative melting.

Fire Climax

Monk (1966) in dry areas of Florida suggests evergreen leaves as an adaptation for cycling through continuous fall, drip, and other characteristics. Some, but not all, dry systems develop storages of organic matter that accumulate until they are mineralized by fire. The adaptation of plant structure to permit this and to provide driable and inflammable substrates provides a means for diverting the potential energies of the sun's heating into drying so that the substrates may support fire regeneration, as in the Southeastern U. S.

Succulents

The well known metabolic day-night inversion of gas exchange in succulents such as cactus seems to be another case of delayed transients. Carbon dioxide is fixed and oxygen released at night. For example, Ting and Dugger (1968) describe intermediates in the biochemical processing so that light fixed during the day is stored in chemical potential energy of organic acids (malic acid) without full processing until night. Malic storages decay faster in light. Recognized already as a water conservation mechanism, the possibilities of an accelerating photorespiration role may be considered when outside nutrients and water are limiting and only the internal cycle is available. Cactus chlorophyll is very high when water is abundant and much less when conditions are dry outside.

Autumn Leaves

In the temperate forest in autumn and in tropical forests at the time of the dry season, the leaves turn red and yellow as chlorophyll is removed and other pigments remain. Where light is brighter, the color seems brighter and develops sooner. Are these pigments serving to accelerate reduction, catabolism, and translocation of valuable nutrients prior to leaf fall? In bright light where photosynthesis has accumulated more photo-synthates, a program of reduction may be greater.

Many shade plants such as aquatic plants or tropical forest bromeliads turn red when placed in the sun. With their load of utilization of pho-tosynthate fixed by genetic characteristics, excess light may unbalance the ratio of photosynthesis to respiration. Decreasing chlorophyll and increasing other pigments may keep the system's internal cycles operable and more nearly balanced.

Red Algae on Snow

On the surface of old snow there may be growths of red algae. In this stressed environment, the pigments may serve to accelerate the recycling by utilization of light. At least there is heating through thermal absorp-tion. Are there more specific energy pathways of photorespiration? Fogg (1967) found 4500 cells per cc with predominance of carotenoid pigments and low rates of apparent photosynthesis.

New Yellow Leaves in Climax Forests

The new growing leaves of rain forests and temperate deciduous forests emerge yellow with pigments prior to addition of chlorophyll. At this stage respiration is in excess of photosynthesis, and organic substrates are being translocated in from other parts of the tree for growth. Is it possible that photorespiration is a predominant process at this stage, the extra energy of the sun coupled to help with the growing processes?

Lignin Derivatives in Waters

Heavily colored swamp waters and derivatives from paper mills have aromatic compound fractions of lignin breakdown that are highly colored, absorbing bright lights in tropical waters such as the Amazon. Many such waters, including the Santa Fe River in Florida, have very high respiration, their oxygen level holding half saturation with a balance between oxygen diffusing in and consumption of oxygen by the aquatic system. Sometimes experiments on oxygen change in light and dark bottles show greater respiration in the light bottles than in the dark ones. If through temperature elevation or other means, some photochemical boost to consumption is involved either directly or through bacterial actions, dark swamp waters may qualify for another case of photoregeneration.

Primitive or Specialized Roles

Since photorespiration seems so widely distributed in many kinds of communities that seem primitive, is it possible that this was a more prevalent process in the past? How many of our fossil plants, and even animals, derived aid from adaptive coupling of light to consumptive and growth processes? Was the origin of life a coupling of two photic processes, photopolymerization and photodissociation?

Implication for Urban Man

If the many special adaptations for diverting energies into regenerative recycling and reorganization of natural systems is a part of effective persistence and survival, some parallel may be seen in man's urban culture, which is in danger of becoming clogged with accumulations, wastes, old structure, and other manufactured units of our system. The instinctive tendencies of some youth to be destructive may be serving now as in the past as part of a probing for the emerging self design process, stabilizing man's system by developing specialist modules for improving the cycling and reordering of our cities and other institutions.

The decaying cores of many of our largest urban centers stand as mute evidence of the need for continual regeneration and regrowth. Man can choose to channel his energies in this direction through uncontrolled fires, riots, and continual decays or through the best efforts of controlled urban renewal and social concern.

Acknowledgment

Dr. A. Moore, Lawrence Burns, and Robert Kelley aided with simulation and scaling of the operational analogs.

Literature Cited

Brock, T. D. and M. L. Brock. 1969. Effect of light intensity on photosynthesis by thermal algae adapted to natural and reduced sunlight. Limnology and Oceanography, 14:334-341.

Bunt, J. S. 1969. Observations on photoheterotrophy in a marine diatom. J. Phycol. 5:37-42.

Caperon, J. 1967. Population growth in micro-organisms limited by food supply. Ecology. 48:709-889.

DiSalvo, Louis H. 1969. Regenerative functions and microbial ecology of coral reefs. Ph.D. dissertation, Univ. of N. C., Chapel Hill.

Emerson, R. and C. M. Lewis. 1943. The dependence of the quantum yield of *Chlorella* photosynthesis on wave length of light. American Journal of Botany, 30:165-178.

Eppley, R. W. and F. M. Macias. 1963. Role of the alga *Chlamydomonas Mundana* in anaerobic waste stabilization lagoons. Limnol. and Oceanogr. 8:411-416.

Fogg, G. E. 1965. Algal culture and phytoplankton ecology. Univ. of Wisconsin Press, Madison and Milwaukee. 126 pp.

—————. 1967. Observation on the snow algae of the south orkney islands. Philosophical Trans. Royal Society Series B 252:279-287.

French, C. S. and D. C. Fork. 1961. Computer solutions for photosynthesis rates from a two-pigment model. Biophys. J., 1:669-681.

Gessner, F. 1937. Untersuchungen uber Assimilation und Atmung submerser Wasserpflanzen. Jb. wiss. Bot. 85:267-328.

Henrici, A. T. 1939. Problems of lake biology. Amer. Assoc. Advanc. Sci. Sympos. 10:39-64.

Kingsbury, J. M. 1956. On pigment changes and growth in the blue green alga. *Plectonema nostocorum* Bornet ex Gomont. Biol. Bull., 110:310-319.

Kowallik, W. 1967. Chlorophyll-independent photochemistry in algae pp. 467-477. *in* Energy conversion by the photosynthetic apparatus. Brookhaven Symposia in Biology, No. 19, 514 pp.

Lugo, A. E. 1969. Energy, water, and carbon budgets of a granite outcrop community. Ph.D. dissertation, Univ. of N. C.

Manny, B. A. 1969. The relationship between organic nitrogen and the carotenoid to chlorophyll-A ratio in five fresh water phytoplankton species. Limnol. and Oceanogr., 14:69-79.

Margalef, R. 1959. Pigmentos assimiladores extraidos de los colonias de celenteros de los arrecifes de coral y su significado. Ecologico Invest. Pesquera, 15:81-101.

—————. 1965. Ecological correlation and the relationship between primary productivity and community structure. Mem. Ist. Ital. Idrobiol., 18:355-364.

Milsum. 1966. Biological control systems analysis. McGraw Hill Book Co., New York, 466 pp.

Monk, C. D. 1966. An ecological significance of evergreenness. Ecology, 47:b.504-505.

Nixon, S. 1969. Characteristics of some hypersaline ecosystems. Ph.D. dissertation, Univ. of N. C.

Nixon, S. and H. T. Odum. 1970. A model for photoregeneration in brines. ESE Notes, 7 (i): 1-3.

————————. 1957. Trophic structure and productivity of Silver Springs, Fla. Ecology Monograph, 27:55-112.

Odum, H. T. 1967. Biological circuits and the marine systems of Texas. pp. 99-157 *in* T. A. Olson and F. J. Burgess, (eds.) Pollution and marine ecology. Interscience.

————————. R. J. Beyers and N. E. Armstrong. 1963. Consequences of small storage capacity in nannoplankton pertinent to measurement of primary production in tropical waters. J. Mar. Res., 21(3):191-198.

———————— and G. Cintrón. 1970. Forest chlorophyll and radiation Chapt. 3 *in* H. T. Odum and R. F. Pigeon, (eds.) A tropical rain forest. Div. of Techn. Information, U.S. Atomic Energy Commission, (in press).

———————— and C. M. Hoskin. 1957. Metabolism of a laboratory stream microcosm. Publ. Inst. Mar. Science, 4(2):115-133.

————————, A. Lugo, and L. Burns. 1969. Metabolism of forest floor microcosms, Chap. I-3 *in* H. T. Odum and R. F. Pigeon, (eds.), A tropical rain forest.

————————, W. McConnell, and W. Abbott. 1958. The Chlorophyll "A" of Communities. Pub. Inst. Mar. Sci., Univ. Texas, 5:65-96.

———————— and R. F. Pigeon. 1969. A tropical rain forest. Div. of Technical Information and Education, Oak Ridge, Tenn. (in press).

———————— and R. C. Pinkerton. 1955. Time's speed regulator, the optimum efficiency for maximum power output in physical and biological systems. American Scientist, 43:331-343.

————————, R. Cuzon, R. J. Beyers and C. Allbaugh. 1963. Diurnal metabolism, total phosphorus, Ohle anaomaly, and zooplankton diversity of abnormal marine ecosystems of Texas. Publ. Inst. Mar. Sci. 9:404-453.

————————, W. L. Siler, R. J. Beyers, and N. Armstrong. 1963. Experiments with engineering of marine ecosystems. Publ. Inst. Mar. Sci., 9:373-403.

Phillips, J. N. and J. Myers. 1954. Measurement of algal growth under controlled steady state conditions. Plant Physiology, 29:148-152; 152-161.

Sollins, P. 1969. Measurements and simulation of oxygen flows and storages in a laboratory blue-green algal mat ecosystem. M.A. thesis, Univ. of N. C.

Steeman-Nielsen, E. 1962. Inactivation of the photochemical mechanism in photosynthesis as a means to protect the cells against too high light intensities. Physiologia Plantarum, 15:161-171.

Ting, I. P. and W. M. Dugger. 1968. Non-autotrophic carbon dioxide metabolism in cacti. Bot. Cag. 129:9-15.

Wassink, H. C., S. D. Richardson and G. A. Preters. 1956. Photometric adaptation to light intensity in leaves of *Acer pseudoptatanus*. Acta Bot. Neerlandica, 5:247-256.

Zelitch, I. 1967. Water and CO_2 transport in the photosynthetic process *in* A. S. Pietre and F. A. Greer and T. J. Army (eds.), Harvesting the Sun in Photosynthesis in Plant Life P. 231-248. Academic Press, N. Y., 342 pp.

George W. Saunders

Carbon Flow in the Aquatic System

Abstract

Available information on the distribution and activities of micro-organisms in the sea and in inland waters indicates that these two variables are continually fluctuating during the annual period, i.e., they are never in steady state. It is known for one Michigan lake that the annual cycles of phytoplankton, bacteria, zooplankton, soluble carbohydrate, soluble organic nitrogen, and soluble organic carbon are different and are not in phase. This indicates that both the quality and quantity of the micro-organisms vary spatially and temporally within aquatic systems. This is true in the tropics as well as north temperate regions and therefore is probably a world-wide phenomenon.

It is possible to estimate the initial conditions for carbon concentration of all the general forms of carbon and most of the rates of carbon transport in certain aquatic subsystems. It appears that changes in the structure of the carbon flow system are controlled by the microorganisms and not directly by the environment, i.e., microorganisms have considerable capacity to adjust their responses to changes in environmental conditions.

Descriptions of longer term fluctuations in distributions of micro-organisms and their activities, as well as descriptions of daily or semi-daily rates of carbon transport, cannot reveal the control mechanisms that are operating to determine the density structure or the functional structure of the microbial component of the aquatic system.

Generalized data from field analyses are presented to show that light is a fluctuating forcing variable within the 24-hour period. The photosynthetic rate responds with a fluctuation not directly proportional to fluctuating light intensity, and both secretion rate of soluble organic matter and bacterial assimilation rate of this secreted organic material fluctuate during the 24-hour period. These latter two fluctuations are not in phase with one another or with the photosynthetic rate fluctuation. This sequence of forcing variable and non-phased fluctuating responses can be viewed as a linearly-coupled control system. Another control system that exists is that of light-forced photosynthesis and the light-cued zooplankton migration which results in a coupled producer-grazer control system. It is suggested that changes in the composition of algae under simulated day-night conditions reflect an active internal control system for these organisms and infer that possibility for other organisms.

An aquatic subsystem, and in fact the whole aquatic system, can be viewed as a set of oscillating control systems having a period of 24 hours, in addition to other longer or shorter period fluctuations. If a sequence of 24-hour oscillating sets is examined, the pathway by which a subsystem moves through a series of states can be described. If the subsystem is subjected to certain critical minor environmental perturbations and the responses of the system observed, it may be possible to reveal where the fluctuations are tightly coupled and the nature of the control mechanisms operating in the system and dictating the manner in which that system passes through a certain sequence of states.

Discussion

Structure and function usually are very closely related, but the knowledge of one does not necessarily reveal the exact nature of the other. Structure in the aquatic system has been examined in many ways but traditionally it has been integrated as energy flow on an annual basis in trophic levels within the system (Juday, 1940; Lindeman, 1942; Dineen, 1953; Odum, 1957; Teal, 1957; Nauwerck, 1963). Because of the difficulty in treating the bacteria they have usually been omitted from the budget. Any analysis to be developed from annual data usually must apply to some degree of sub-geological time and therefore to some aspect of lake evolution. Since the annual period is greater than the longevity of most microorganisms, the historical record of the sequence of microbial communities is lost in the integration or, in fact, by never having been examined. Therefore it becomes impossible to make any specific statement about the structure or the role of microbial communities within shorter time intervals. Many of the questions we purport to be asking require that analyses be conducted within time intervals of hours or days rather than years.

I want to describe the results of microbial activities and the structure of microbial processes in subsystems of the aquatic system within short intervals of time. I want also to indicate how short term measurements might be useful in revealing function and may lead to an analysis of mechanisms of control and regulation within the system. I want to do this from the point of view of carbon flow, recognizing that energy flow and nutrient cycling are absolutely coupled, and that to discuss one without the other is merely for convenience. The cycling of carbon is a vehicle for the transport of energy along one pathway in the dissipation of energy from the ecosystem.

When one examines the distribution of nutrients and the metabolic products of organisms during the period of one year in lakes and oceans (Duursma, 1961; Belser, 1959; Brehm, 1967; Allen, 1969) it is striking that the concentrations of such substances vary during the year. In a detailed study of a Michigan lake (Saunders and Lauff, unpublished) the annual concentration distributions, in the lake volume, of soluble carbohydrate, soluble organic nitrogen, soluble organic carbon, phytoplankton, and bacteria, are very different. Presumably these fluctuations reflect changes in the quality and quantity of microorganisms as well as of their metabolic activities. Even in the tropics where light and temperature are relatively more constant, major fluctuations in concentration of organisms or their metabolites are observed. The structure, qualitative and quantitative, and the activity of the microbial community does not appear to be in steady state during this interval of time. Where data are available on the time-space distribution of phytoplankton, bacteria, and zooplankton (Overbeck, 1968; Gambaryan, 1968; Rasumov, 1962; Saunders, 1963) it is possible to generate some general correlations among these groups of organisms, which possibility suggests that they are interacting and exerting some kind of control over one another. However, it is difficult to extract any more specific correlation which would reveal the true mechanisms of interaction.

In annual studies of two Michigan lakes (Saunders, unpublished) only three points in time were observed where phytoplankton and bacteria were specifically correlated in terms of changes in their populations' densities. In the single case where a zooplankton increase and a phytoplankton-bacterial minimum were observed, subsequent data suggest that grazing by zooplankton was not the cause of the phytoplankton-bacterial minimum. Therefore another factor or other factors must have been operating to determine the simultaneous changes of the phytoplankton and bacteria. The simple correlation observed did not reveal the true control mechanism for the phytoplankton and bacterial concentrations. During the remainder of the year, which constitutes 90% of the period, no obvious direct correlations between phytoplankton-soluble organic matter-bacteria were observed. This suggests either that there were no direct relationships or that the analytical procedures were such that they could not reveal any specific relationships. In earlier investigations of the time course of bacterial numbers in natural waters, it has not been possible to predict their distribution during the year with any confidence. The season of maximum development differs among lakes, and within a lake may vary from year to year (Henrici, 1939; Taylor, 1940; Rasumov, 1962).

It is also known that the vertical distribution of bacteria in stratified lakes can change very rapidly (Rasumov, 1962). An example of short term changes in the vertical distribution of bacteria in Frains Lake is presented in Fig. 1. The most drastic changes occurred in mid-June following a pulse of phytoplankton. At this time the numbers of bacteria

FRAIN'S LAKE 1959-'60

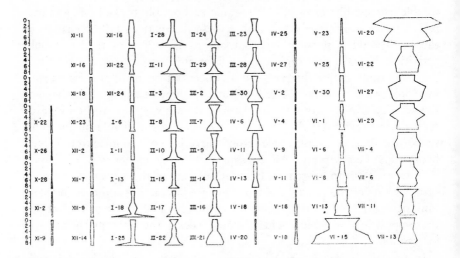

FIGURE 1. Relative vertical distribution of bacteria (direct counts) by date in Frains Lake, Michigan. (Ordinate: depth in meters; abscissa: relative count).

increased sharply and decreased again in a few days. The distribution of bacteria inverted and reverted in this same time interval. This indicates that there are shorter term variables and processes that control the distribution of bacteria and that are not essentially revealed by the more general correlations discussed in the preceding paragraph.

In ecology we have long discussed the Law of the Minimum. In nature, however, it may be only rarely that a population responds to a single environmental variable. This would occur when the population has been stressed in the extreme and driven toward the boundaries of its possible existence. At this time a specific correlation would easily be observed. At other times, which probably constitute the majority, any population may be responding to a set of variables (McCombie, 1953, 1960). It becomes difficult to assess the impact of the separate variables of the set and also to determine the true response of the population. The analytical procedure needed for meaningful investigation must have sufficient sensitivity to resolve changes in the operating variables and their impact as well as the qualitative and quantitative responses of the organisms. The frequency of sampling must lie well within the longevity of the organisms and, in fact, well within the period of the event, if we are to discern the pathway by which the system moves from its initial to final state and the mechanisms that determine this pathway.

It is necessary to conclude that we have been examining the microbial component of ecosystems in a manner that is inappropriate for revealing microbial function and the mechanisms that control and regulate this function.

I would suggest that if we are to answer contemporary questions concerning the microbial component of ecosystems, it is necessary to discontinue casual examination of the system and it is absolutely necessary to examine the fine structure of the system and its perturbations. This is a procedure which has been well used with profit in other fields. It is also necessary to conduct the analysis in a natural system which is minimally perturbed by the manipulation. While laboratory experiments can be used to answer more general questions it is also well documented that extrapolation of results from reconstructed systems is not always reliable.

I would like in the following to indicate that it is feasible to estimate the concentration structure and the metabolic structure of a microbial community. It should also be possible to analyze the perturbations of environmental variables and the response of the community to those perturbations. It is also possible to do this within very short time periods, including parts of a day.

If a set of variables operating on a system changes, the microbiota will respond to the changes in three possible ways: (1) the organisms will respond passively, in which case the organisms are caused to change to some new state where the change is controlled directly by the environment; (2) the organisms will regulate, i.e., they will attempt to maintain constancy and the changes observed will merely reflect the degree to which they are unable to do so; (3) the organisms will control their change of state, i.e., they will dictate the exact path they take in proceeding to the new state. I suggest that it is the last which is in fact the case (there is abundant evidence for regulation and control at the molecular level and the same is intuitively obvious at other levels of organization in the hierarchy of life structures).

It is possible to estimate the initial conditions at sunrise and the rates of carbon flow in the aquatic system on a 24-hour basis for all the general forms of carbon (Fig. 2). The analysis was made on a subsystem of lake water from a single depth in the photic zone of Frains Lake. Radioactive tracers were used as an analytical tool and the experiment summarized is the first of a series designed to test the feasibility of such an approach. The figure implies a steady state condition for the rates during the period. One point is very apparent. The carbon cycling between the algal and inorganic forms exceeds the carbon being cycled through all the other forms of carbon in this subsystem.

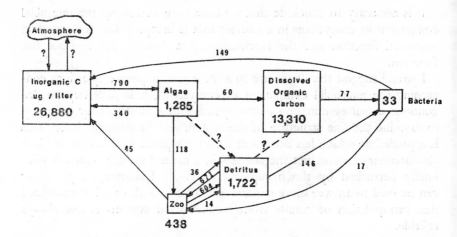

FIGURE 2. Carbon flow in microbial system at 1.0 meter in Frains Lake, Michigan 22-vii-68. (Numbers within blocks: carbon concentration at sunrise in ugC liter; numbers on arrows: transfer rates in ugC per liter per 24 hours).

The structure of the carbon system is summarized in Fig. 3. The upper bar graphs represent the concentration structure of the system. There are two very large pools of non-living carbon, the inorganic and the dissolved organic carbon. Both of these forms are very much in excess of the apparent requirements of the living organisms. The algae and detritus (probably planktogenic) are of intermediate concentration and the bacteria and zooplankton (consumer organisms) have the lowest concentrations. The bacteria amount to about 10% of the zooplankton concentration. This pyramid of carbon distribution probably should not be unexpected for the upper photic zone in lakes. It is unique that all the forms of carbon have been estimated and included in the budget.

The metabolic structure of the subsystem is presented in the lower part of Fig. 3. These bar graphs represent the input rates or the gross assimilation rates of carbon in micrograms of carbon per liter of lake water per day for each form of carbon. It becomes obvious from the figure that the microbiota control the movement of carbon in the system even though the non-living forms of carbon are by far the dominant ones in terms of biomass. This implies that although the intensity of sets of environmental variables will set the general level of operation of the system, the chemical processes of life have some freedom of expression and will control the modification of the initial environmental distributions and will dictate their own performance and distributions within that environment.

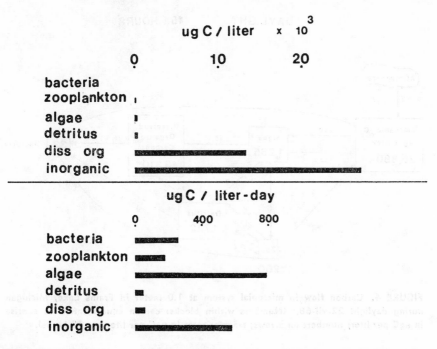

FIGURE 3. Structure of microbial system at 1.0 meter in Frains Lake, Michigan 22-vii-68. (Upper bar graphs: initial concentration structure at sunrise; lower bar graphs: metabolic structure as input rate in ugC per liter per day).

The statement of initial conditions and the daily rates of transfer of various forms of carbon may be somewhat misleading, because within a 24-hour period the amounts of carbon species are not constant nor are the rates of transfer. The 24-hour period can be partitioned into at least two periods, day and night. The initial conditions and the total transfer rates for the daylight period are presented in Fig. 4. The concentration structure is not changed fundamentally but the metabolic structure is shifted toward a greater relative importance of cycling between the algae and inorganic carbon than in Fig. 2. This should obviously be true since photosynthesis can occur only during daylight whereas hetero-trophic activity occurs throughout the day. The initial conditions and the rates of transfer for the nighttime period are shown in Fig. 5. The concentration structure is somewhat modified due to changes during the day. The initial conditions for the zooplankton include those organisms which have undergone vertical migration into the depth layer of the sub-system studied as well as growth within the system during the day. The

DAYLIGHT   15 HOURS

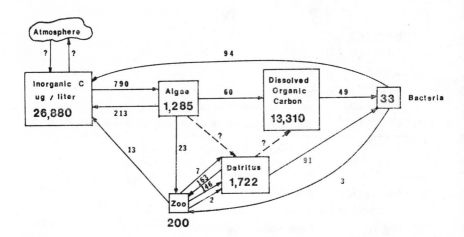

FIGURE 4. Carbon flow in microbial system at 1.0 meter in Frains Lake, Michigan during daylight 22-vii-68. (Numbers within blocks: carbon concentration at sunrise in ugC per liter; numbers on arrows; transfer rates in ugC per liter per 15 hours).

NIGHT   9 HOURS

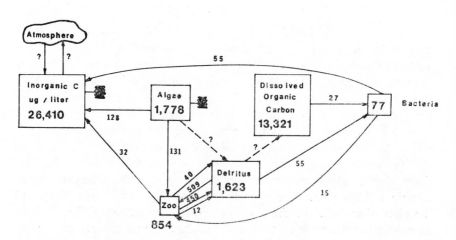

FIGURE 5. Carbon flow in microbial system at 1.0 meter in Frains Lake, Michigan during nighttime 22-vii-68. (Numbers within blocks: initial carbon concentration at sunset in ugC per liter; numbers on arrows: transfer rates in ugC per liter per 9 hours).

metabolic structure of the subsystem has changed completely. Algal photosynthesis and algal secretion of soluble organic matter have gone to zero rate. The system is completely heterotrophic and the zooplankton have come to dominate metabolism in the system. This is because there has been immigration of zooplankton into the depth volume studied, which occurs at the onset of darkness. This complete shift in the metabolic structure of the system is quite remarkable. It results, however, only in more minor changes in the concentration structure.

It is, however, more interesting and more realistic to examine the time course of concentrations and rates continuously within a 24-hour period. The data points for the time course of cumulative net particulate photosynthetic carbon fixation and of cumulative carbon secretion for a sample of lake water are given in Fig. 6. There are two points that need to be made. Firstly, there are inflections in the curves for particulate fixation and secretion. This means that the rates go from zero to some high value to zero, i.e., the rates fluctuate. Secondly, the points of inflection are not in phase so that the secretion rate curve lags the photosynthetic rate curve by about two hours, in this case.

This kind of data is generalized in the set of schematic diagrams given in Fig. 7 where the time interval is 24 hours and the initial time is sunrise. Light intensity varies with the sun's angle during the day. The phytoplankton respond to changes in light intensity but the response is not directly proportional. The photosynthetic rate per unit volume is asymmetric about mid-solar day, being depressed in the afternoon. This suggests that the phytoplankton may have some control over the manner in which they respond to light. The phytoplankton secrete photosynthate in which the secretion rate is out of phase with the rate of photosynthesis. In an independent experiment I have been able to demonstrate that the assimilation rate of secreted material by the indigenous bacterial community of a lake water sample fluctuates and is out of phase with the secretion rate, in this case by approximately 6 hours. It is possible to view this sequence of events as representative of a tightly coupled reaction system in which light intensity, in its ordinary diurnal variation, is a forcing environmental variable. The biota, phytoplankton and bacteria, respond to this perturbation in a sequence of fluctuating causal reactions which presumably are under the control of the organisms. A part of the soluble organic pool intervening the phytoplankton and bacteria is a component of this sequence.

Another coupled reaction system is that of the primary producers and the herbivorous zooplankton grazers. This can in fact be viewed as two systems. In one system the biomass of phytoplankton fluctuates during

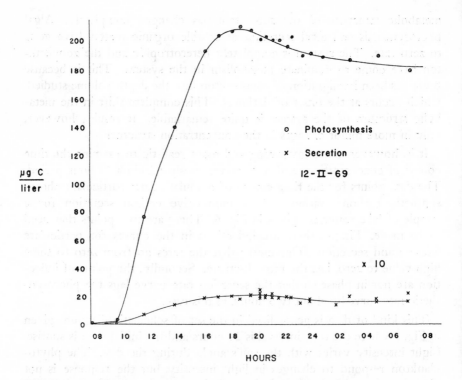

FIGURE 6. Cumulative carbon in particulate and soluble phases due to photosynthesis during 24 hours in Frains Lake, Michigan at surface under ice. Sunrise 0730.

the day as a function of light intensity, daytime grazing rate, respiration, night-time grazing rate, as well as natural mortality rate. The curve for cumulative phytoplankton carbon fixation (Fig. 6) is representative of this, where the initial biomass is not zero. The grazing rate is a function of diurnal migration of the zooplankton which is cued by changes in light intensity or rate of change in light intensity. The second system has to do with assimilation of algae by zooplankton. I have some preliminary evidence that the assimilation rate of algae by migrating *Daphnia* fluctuates during the 24-hour period, being greater during the night than during the day.

In addition to the diel fluctuations of such gross external and internal variables, it is well known that cultures of algae undergoing sequential light and dark exposures exhibit marked changes in the proportions of their biochemical constituents. A schematic representation of this is given for the time course of protein and carbohydrate in percentage in Fig. 7.

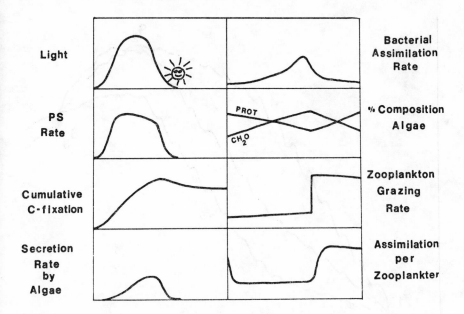

FIGURE 7. Schematic diagrams of the distribution of concentration or rates of processes from sunrise to sunrise in the upper photic zone of an aquatic system.

These and other fluctuations in biochemical constituents fall generally within the realm of regulation and control at the molecular level.

The above general considerations can be extrapolated broadly to the ecosystem so that it may be viewed as a network or a set of connected fluctuating reaction systems, in which the fluctuations vary in shape from quasi-sinusoidal to square waves or paired step functions and in which the fluctuations are not in phase and may be more or less damped. The change of state of any system over a unit time interval may be regarded as the summation of this network of fluctuations.

If the state of an ecosystem is known at S_1 and S_2 (Fig. 8), there is no way of knowing from that information alone how the system moved from State-1 to State-2. It could have moved along an infinite number of pathways. If the system is fluctuating in shorter time intervals with a more or less constant period and if one samples as frequently as the period of fluctuation, the path taken in moving from State-1 to State-2 can be described crudely. If the system is examined more frequently than the period of fluctuation, the path taken from State-1 to State-2 can be described more precisely. However such information will not reveal the functional mechanisms which caused the system to move from State-1 to State-2. The mechanisms can only be discerned if the causal operators

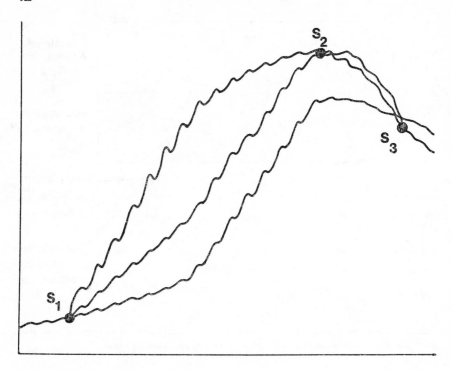

FIGURE 8. The state of a fluctuating integrated reaction system with time. (Ordinate; $S = f(r,s,t, \ldots \ldots z)$; abscissa: time).

on the system can be revealed, Fig. 9. If one is clever enough to guess the major environmental operators, perturbations of these operators can be performed and the response of the biota and the rest of the system determined. In such a manner the major causal reaction systems can be revealed and presumably also the reaction mechanism.

The pelagic region of a lake is constituted mainly of microbial or microscopic communities, in which I include the zooplankton. Such being the case, it is necessary to examine this system within the lifetime of the populations and more importantly within the critical response time of the populations. Otherwise the historical record of the perturbation-response reaction is lost, never having been examined. Microorganisms can respond to environmental changes in a matter of seconds or minutes. This has become obvious in terms of induction-repression mechanisms. The basis for these mechanisms lies at the molecular level. If the time constant for the molecular mechanisms is very short relative to population responses, perhaps the molecular mechanisms can be considered as para-

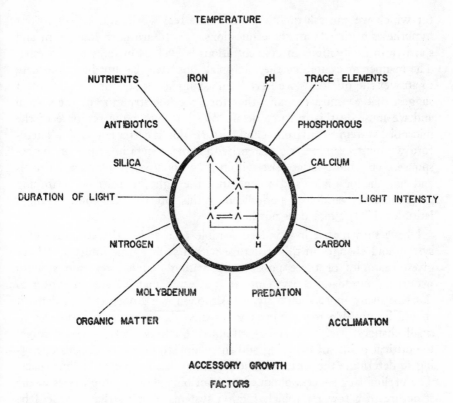

FIGURE 9. Schematic diagram of a set of environmental factors operating on an energy flow system at an instant in time.

meters for longer term reaction systems. If the 24-hour period is considered as a pragmatic compromise, it has a periodic fluctuation in light which is constant and nearly symmetrical for many latitudes; it is not much longer than the lifetime of some microbes and certainly much shorter than the longevity of most microbial populations. Therefore if we examine perturbations in the fine structure of the environment and in fine structure of the ecosystem or a subsystem within a 24-hour period, it should be possible to analyze function and control within that system during this period. If a sequence of 24-hour analyses is made, it should be possible to perform something analogous to a power spectrum analysis of observed fluctuations. If a residuum occurs after short term interactions have been removed, this should reveal longer term interactions and suggest mechanisms of control operating during these longer periods.

The preceding discussion indicates that aquatic ecosystems are dynamic. They are probably never in steady state, at least within time intervals for which we can ask questions about the real world and expect to test hypotheses arising from these questions. There are both long term and short term fluctuations in concentrations as well as in rates of processes. The manner of examining the microbial subsystem has not been adequate to answer the questions we have been asking about aquatic ecosystems. I suggest that we must examine, therefore, the fine structure of the system and we must do this in some detail. We know that the response of the microbial system to environmental perturbation is extremely rapid. Therefore we must examine this system within the generation time of the response. This must lie at least within a 24-hour period at most latitudes and has the pragmatic advantage that the light intensity as a forcing environmental variable is periodic over this interval. Of course at high latitudes this last fact does not hold.

I have shown that it is indeed feasible to examine the structure of a system and changes in this structure within such a time interval. I have given examples of three possible fluctuating coupled reaction systems occurring in aquatic ecosystems. It is inferred that we should expect to describe many more such systems. Although it has not yet been done, it should be possible to examine the response of such reaction systems to small changes in environmental variables. Such an analysis would reveal the critical points of coupling and the major control mechanisms operating to determine the nature of the system at various points in time-space. The preliminary success obtained in describing short term changes in the structure of a few microbial reaction systems suggests that it would be possible to accomplish this last objective also with success. Thus we can visualize the initial phases in the analysis of total integration within the aquatic ecosystem, at least within the limnetic zone.

Acknowledgments

The work on Frains Lake was supported in part on AEC Contract No. AT(11-1)-781, Michigan Memorial Phoenix Project Grants No. 192 and 267, and by a grant from the Louis W. and Maud Hill Family Foundation. I would like to thank Dr. G. H. Lauff and Dr. D. C. Chandler for their help and Dr. Lauff for provision of facilities and supplies while in residence at the W. K. Kellogg Biological Station during the summer 1969. The work was carried out with the help of S. Sanders, L. S. Chertkov, M. S. Misch, D. Brubaker, J. C. Roth and T. A. Storch. I have had the benefit of discussing the work with all of these persons as well as detailed and lengthy discussions of certain aspects of the general problem with Dr. D. Woodring. I would like to thank Dr. C. L. Schelske and J. C. Roth for critical readings of the manuscript.

Literature Cited

Allen, H. L. 1969. Chemo-organotrophic utilization of dissolved organic compounds by planktic algae and bacteria in a pond. Int. Revue ges. Hydrobiol., 54: 1-33.

Belser, W. L. 1959. Bioassay of organic micronutrients in the sea. Proc. Nat. Acad. Sci., 45: 1533-1542.

Brehm, J. 1967. Untersuchungen über den Aminosäure-Haushalt holsteinischer Gewässer, insbesondere des Pluss-sees. Arch. f. Hydrobiol. Suppl. 32: 313-435.

Dineen, C. F. 1953. An ecological study of a Minnesota pond. Amer. Midl. Nat., 50: 349-376.

Duursma, E. K. 1961. Dissolved organic carbon, nitrogen and phosphorous in the sea. Neth. J. Sea Res., 1: 1-148.

Gambaryan, M. E. 1968. Mikrobiologicheskie issledovaniya Ozera Sevan (Microbiological studies of Lake Sevan). AN SSSR 165 pp.

Henrici, A. T. 1939. The distribution of bacteria in lakes. In Problems of lake biology. AAAS Pub. 10, pp. 39-64.

Juday, C. 1940. The annual energy budget of an inland lake. Ecology 21: 438-450.

Lindeman, R. L. 1942. The trophic dynamic aspect of ecology. Ecology 23: 399-418.

McCombie, A. M. 1953. Factors influencing the growth of phytophankton. J. Fish. Res. Bd. Can. 10: 253-280.

——————. 1960. Actions and interactions of temperature, light intensity and nutrient concentration on the growth of the green alga, *Chlamydomonas reinhardii* Dangeard. J. Fish. Res. Bd. Can., 17: 871-894.

Nauwerck, A. 1963. Die Beziehungen zwischen Zooplankton und Phytoplankton im See Erken. Symb. Bot. Upsal., 17: 1-163.

Odum, H. T. 1957. Trophic structure and productivity of Silver Springs, Florida. Ecol. Monogr., 27: 55-112.

Overbeck, J. 1968. Prinzipielles zum Vorkommen der Bakterien im See. Mitt. Internat. Verein. Theor. Angew. Limnol., 14: 134-144.

Rasumov, A. S. 1962. Mikrobial'ny plankton vody (The microbial plankton of water). Trud. vsesoy. Gidrobiol. obshch., 12: 60-190.

Saunders, G. W. 1963. The biological characteristics of freshwater. Proc. Great Lakes Res. Conf., GLRD Pub. No. 10: 245-257.

Taylor, C. B. 1940. Bacteriology of freshwater, I. Distribution of bacteria in English lakes. J. Hyg., 40: 616-640.

Teal, J. M. 1957. Community metabolism in a temperate cold spring. Ecol. Monogr., 27: 283-301.

G. Dennis Cooke

Aquatic Laboratory Microsystems and Communities

Abstract

Aquatic laboratory ecosystems and communities are described. Their activities closely parallel similar but larger units of nature. The microcosm method involves isolation of representative samples of an ecosystem, development of a "stock" system, and establishment of many replicate experimental ecosystems under the desired physicochemical conditions. Microcosms have a diurnal metabolism like that of natural systems; causal factors in this pattern of metabolism are discussed. Ecological succession, the unifying theory in ecology, has been studied in laboratory systems; these data, and some internal regulatory activities of mature microecosystems, are described. The principles of succession, as studied in laboratory ecosystems, are discussed in relation to the development of life support microecosystems for space travel. Laboratory stream communities have been employed in comparative analyses of metabolism measurement methods and in studies of short-term succession. The pH, O_2, substrate, and chlorophyll methods are discussed. Variations in species composition, biomass levels, and stream community metabolism under controlled conditions of temperature, light, and current velocity are compared.

These laboratory systems are important ecological tools for investigating the properties of ecosystems and communities, and may become important for the testing of potential pollutants on levels of biological organization to which they are applied in practice.

Introduction

The analysis and understanding of large units of nature are among the major objectives of ecology. The overall picture of the structure and function of communities and ecosystems has come largely from the summation and integration of many separate investigations on smaller segments of these systems. The major difficulties, however, in this approach to the understanding of communities and ecosystems are (1) the large size of such units so that only a very small portion can be studied at any one time, and (2) the almost complete absence of replicability so that few controlled experiments are possible. Recently, a new experimental tool in ecology, the laboratory ecosystem or community, has been successfully employed in ways which have contributed significantly to the development of ecological theory. The main advantages of these systems are that their structure and activities closely parallel much larger natural systems, and that they are replicable and controllable and thus amenable to experimentation. The artificiality of unispecific cultures in ecological studies, a problem thoroughly discussed by McIntire (1969), is avoided and at the same time environmental factors can be regulated so that autecological as well as synecological questions may be asked by the experimenter.

There are two basic types of experimental laboratory systems. One, the closed beaker or flask-type, is an ecosystem (as defined by E. P. Odum, 1963), and is often called a microcosm or microecosystem. The other is the open laboratory community containing both autotrophic and heterotrophic components, and usually an abiotic portion as well, and is part of a larger outside ecosystem. In both, the assemblage of organisms is natural and is usually obtained from nature as a unit. These groups of organisms are allowed to reorganize and reassemble under laboratory conditions in several to many replicate units, so that the investigator has a series of controllable, repeatable, manipulatable, miniature ecosystems or sections of ecosystems.

The object of this paper is to describe the laboratory community or ecosystem as one of the newer experimental tools for the study of these levels of biological organization, and to discuss the results of some recent studies.

Discussion

The Microcosm Method

The use of aquaria or large tanks for research and teaching purposes is well known. In most instances, these have been uncontrolled situations, without replicate units, and successive samples by the investigator progressively altered the systems. Often, the medium was not defined and the vast bulk of organisms was unknown. Beyers (1962a,b; 1963a) was among the first to employ many small replicate laboratory ecosystems, and his technique is representative of what has become known as the "microcosm method."

Beyers' basic laboratory ecosystem was developed from a composite sample of a sewage oxidation pond near Austin, Texas. The material from the pond was placed under a pond-like laboratory regime of temperature and light and then allowed to go through succession. A large "stock" system was thus created.

Two different types of media are used by Beyers. The basic medium is sterile one-half strength Taub and Dollar (1964) #36 solution. If 5 mg. thiamine per liter plus 35 mg. of fixed nitrogen per liter are added to this base medium, the microecosystems will undergo an autotrophic succession when an aliquot of the stock system is added to each sterile flask. Or, if 0.05% proteose-peptone is added instead to the base medium, the new ecosystems will experience a bacterial bloom or a heterotrophic succession, which will then be followed by the autotrophic succession. In either case, the new ecosystems are carefully cross-seeded to minimize any possibility of divergence. Replicate laboratory aquatic ecosystems do not differ significantly with respect to levels of community metabolism (Abbott, 1966).

Depending upon the objectives of the experiment, the light, temperature, and other physical conditions can be varied. In addition to beakers and Erlenmeyer flasks, glass carboys have been used as containers for microcosms (McConnell, 1962, 1965; Abbott, 1966, 1967), Kurihara (1960) has reported on the use of the internodes of bamboo of different ages, and Coler and Gunner (1970) have employed modified disposable tissue culture flasks. The standard light source used by Beyers has been Gro-lux (Sylvania), with a cycle of 12 hours light—12 hours dark. From a few to hundreds of replicate microecosystems are thus established by the investigator under the desired conditions of medium and physical environment. All of the new ecosystems undergo succession at the same rate.

Standard ecological techniques are employed for the analysis of the structural and functional features of the experimental ecosystems. Community photosynthesis and respiration are usually measured by the pH method (Beyers et al., 1963) and at intervals some systems are sacrificed for population counts, biomass, and chlorophyll estimates. One of the most valuable features of the microcosm method is that the measurement method itself will not affect the overall results, since sufficient replicates are established to allow the investigator to sacrifice one or more entire ecosystems at frequent intervals during the study.

Beyers' basic microecosystem is nearly gnotobiotic. The species have been listed by Gorden et al. (1969). The ecosystem has three producer, five consumer, and eleven bacterial species. His microsystem is probably the most complex gnotobiotic ecosystem known to science. Taub (1969) has also developed a gnotobiotic ecosystem consisting of one producer species, one consumer species, and three decomposer species, and Nixon (1969) has described a gnotobiotic brine microecosystem. These last two microecosystems differ from that of Beyers in that the systems were assembled species by species rather than by reorganization of a group of organisms found together in nature. Most other investigators who have used the microcosm method have not used gnotobiotic ecosystems.

The sections to follow describe some results of studies which have employed the microcosm method or some variation of it.

The Ecology of Laboratory Ecosystems

Diurnal Metabolism

One of the most important characteristics of large units of nature is their diurnal variation in metabolism and nutrient cycling, and in structural features such as amount of photosynthetic pigments (e.g., Jackson and McFadden, 1954; Shimada, 1958; Goering, Dugdale, and Menzel, 1964). The metabolic activity of laboratory microecosystems also has a distinct diurnal pattern (Beyers, 1963a,b,c; 1965). Fig. 1, based on the average of 100 diurnal metabolism measurements from 12 benthic microcosms, is typical of the daily variation in metabolism of a microcosm under 12 hours of constant illumination and 12 hours of dark. Just after the lights are turned off, the ecosystems exhibit a burst of respiratory activity, which then declines over the night. At lights on, there is a burst of photosynthesis, which also declines during the remainder of the period. An identical type of metabolism curve was found in a wide variety of microecosystems studied by Beyers (Figs. 2 and 3). The simplest ecosystems, such as the 51°C and the brine systems, exhibit the largest amplitude of variation, while more complex ecosystems, such as the pond type, have less diurnal variation.

Other investigators have reported identical patterns in laboratory microcosms (H. T. Odum et al., 1963; Butler, 1964). Abbott (1966), who used an O_2 method of measuring photosynthesis and respiration, noted that in addition to the day-night pattern there was also a midday depression in photosynthesis which he interpreted to be an endogenous rhythm. Similar diurnal variations in metabolism have not been noted in laboratory stream communities (McIntire and Phinney, 1965), although they have been observed in natural streams (Owens, 1965).

A complete explanation for the shape of the diurnal curve is apparently not possible at this time. Beyers (1963a, 1965), however, has offered several hypotheses. Initially, it was thought that the burst of respiration at the onset of dark was due to the stimulatory effect of the dissolved oxygen which had built-up during the day. But, when dissolved oxygen was held constant by bubbling the microcosm, the burst of respiration,

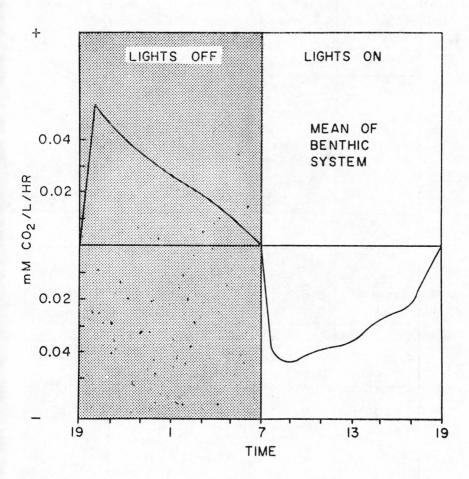

FIGURE 1. Diurnal rates of carbon dioxide uptake and release during net photosynthesis and nighttime respiration in twelve benthic freshwater microcosms. This curve is the mean of one hundred curves. From Beyers (1963b). Reproduced with permission of the author.

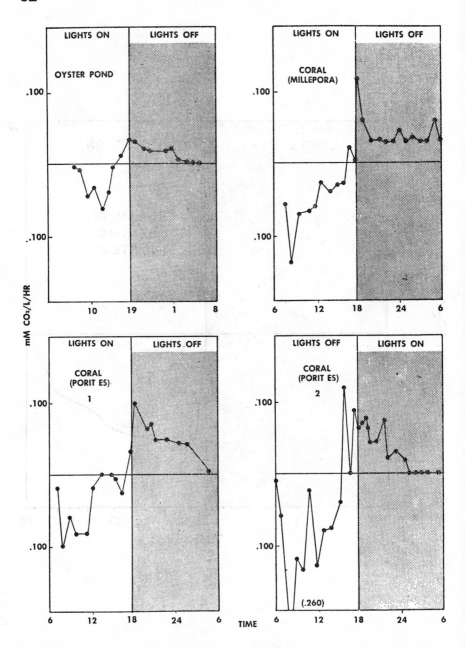

FIGURE 2. Diurnal rates of carbon dioxide uptake and release during net photosynthesis and nighttime respiration in oyster pond microcosm, and three small, isolated coral heads. From Beyers (1963b). Reproduced with permission of the author.

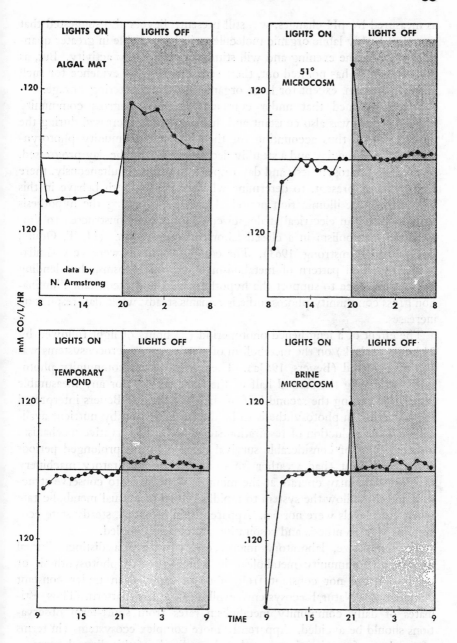

FIGURE 3. Diurnal rates of carbon dioxide uptake and release during net photosynthesis and nighttime respiration in a marine algal mat microcosm, a temporary pond microcosm, and brine microcosm. From Beyers (1963b). Reproduced with permission of the author.

as measured by pH changes, was still present. Beyers also suggested that a pool of highly labile organic molecules will be available in greater quantities early in the evening and will stimulate respiratory activity. But, as Abbott (1966) has pointed out, there seems to be little evidence for such an accumulation, except for large organic molecules entering storage.

Beyers assumed that under constant illumination gross community photosynthesis was also constant and that respiration increased during the lighted period, thus accounting for the high net community photosynthesis early in the day and a steadily decreasing rate as the day progressed. Since gross photosynthesis and day respiration occur simultaneously, there is no way, at present, to determine whether one or both behave in this way during the illumination period. However, assuming the hypothesis to be correct, an electrical analog circuit model was constructed to simulate the metabolism in a closed laboratory ecosystem. (H. T. Odum, Beyers, and Armstrong, 1963). The computer curves were very similar to the observed pattern of metabolism in the microcosms, thus lending indirect evidence to support the hypothesis that under constant illumination gross community photosynthesis remains steady while day respiration increases.

The effect of a lengthened photoperiod (24 hours of light followed by 24 hours of dark) on the metabolism of steady-state microecosystems was also investigated (Beyers, 1963a). There was no net community photosynthesis during the second half of the light period, nor any measurable respiration during the second-half of the dark period. Beyers interpreted the reduction in photosynthesis to be due to limitation by nutrient availability. The reduction of respiration suggested a conservative mechanism in ecosystems of considerable survival value. During prolonged periods of reduced light (bad weather for example) the respiratory machinery of ecosystems may operate at the minimum rate so as to conserve structure and thus allow the system to rapidly return to a usual metabolic rate when light levels were normal. Apparently in balanced, steady-state ecosystems photosynthesis and respiration are closely coupled.

To summarize, laboratory microecosystems have a distinct diurnal pattern of community metabolism in which rates of photosynthesis or respiration are not constant from hour to hour, even under constant illumination. Natural ecosystems exhibit a similar pattern. Thus estimates of daily community metabolism rates from short-term observations should be avoided. Apparently more complex ecosystems (in terms of numbers of species) have a reduced amplitude of this diurnal curve.

Ecological Succession in Microecosystems

Introduction. Succession is a complex process of community development that involves a directional, predictable change in community structure and function which results from the modification of the physical environment by the community, and which culminates in as stable an ecosystem as is biologically possible on the site in question (E. P. Odum, 1963). The principles of ecological succession form the basis of a unifying theory of ecology (Margalef, 1963a,b, 1968; E. P. Odum, 1969), and the study of succession in laboratory microecosystems has contributed substantially to our overall understanding of the process.

As indicated earlier, succession in laboratory ecosystems is initiated by introducing aliquots of a stock ecosystem into a group of flasks or beakers which have the desired medium. In the inorganic medium, supplemented with thiamine, an autotrophic succession occurs. If an organic supplement such as proteose-peptone is added to the medium, a bacterial bloom and a period of heterotrophic activity will precede autotrophic succession. Cross-seeding reduces the possibility of divergence and insures replicability.

Succession in microecosystems is unnatural in the sense that there is no immigration nor emigration, only an input of light energy and an exchange of gases with the atmosphere. Therefore, the species composition does not change except in terms of relative abundance and the number of species in encystment stages. This property of laboratory ecosystems may be particularly advantageous for future work since one could compare ecosystems of varying diversity.

Successional processes and the overall pattern of succession in the flask or beaker-type of laboratory ecosystem is representative of a large-scale unit of nature in which the magnitude of exchange with adjacent systems is negligible when compared to the activities within the system.

The Pattern of Succession. As in larger natural ecosystems, various species gain ascendancy in terms of activity and abundance during succession, and then are replaced or become less active. For example, Cooke (1967) observed that plankton-feeding *Daphnia* were abundant only during the "bloom" stage of succession and then were not active in the systems, whereas ostracods, which are detritus feeders, became more numerous as succession proceeded. Of the 30 genera of algae in the ecosystem only about 10 or 12 were abundant at any one time, and at maturity there was a very stable species composition. Other investigators have made similar observations (Beyers, 1963a; Taub, 1969; Nixon, 1969).

Gorden (1967) and Gorden et al. (1969) clearly demonstrated that not all species in microecosystems were active at the same time, and that some species altered conditions so as to favor other species. Gorden found that all of the bacterial species in his microecosystem (identical to Beyers' basic system) were capable of stimulating the growth of *Chlorella*, the metabolically most important autotroph, by releasing thiamine and carbon dioxide. Figure 4 shows the great enhancement of *Chlorella* population growth when bacteria were present in comparison to growth in various media without bacteria but with thiamine or thiamine and sterile air (CO_2 enrichment). Filtrates containing dissolved organic material from a *Chlorella* culture and a bacterial culture were capable of supporting bacterial growth. Thus species, often taxonomically diverse, interact to bring about succession in microecosystems, just as in nature.

The changes in microecosystem metabolism and biomass were studied by Cooke (1967) during an autotrophic succession and by Gorden (1967) and Gorden et al. (1969) for heterotrophic succession. These events in Cooke's study are summarized in Figs. 5 and 6. During autotrophic succession there is an initial burst of photosynthetic activity, in excess of community respiration, so that biomass accumulates. Metabolism increases to a maximum and then decreases until photosynthesis equals respiration (P/R ratio of about 1) and there is no further increase in biomass. During the first five days of Gorden's heterotrophic succession, respiration exceeded photosynthesis (P/R<1) due to the bacterial bloom; then an autotrophic succession like that shown in Figs. 5 and 6, took place. A mature ecosystem is defined as one in which levels of biomass and metabolism are stable (Cooke, 1967).

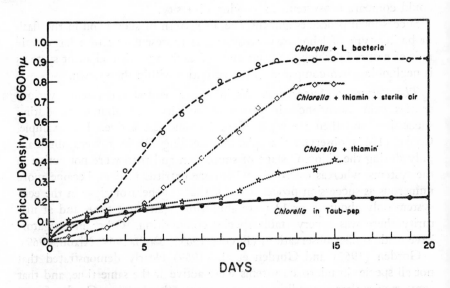

FIGURE 4. Optical density measurements of the growth of *Chlorella* in Taub-pep, Taub-pep plus thiamin, Taub-pep plus thiamine and sterile air, and Taub-pep plus L bacterium. From Gorden et al. 1969. Reproduced with permission of the authors.

Early in succession, when there is a nutrient-induced algal bloom, the ratio of gross community photosynthesis to biomass (P/B) is high (Fig. 6). That is, a small amount of biomass is very productive. As succession proceeds, without any addition of nutrients, this type of efficiency, called ecosystem production efficiency, drops rapidly so that at more mature stages there is a large relatively unproductive amount of organic matter. The relationship between structure and metabolism in ecosystems may be examined in another way. The ratio of biomass to gross community photosynthesis (B/P) is a measure of system maintenance efficiency (Fig. 6). Thus early in succession the ratio is low, but at maturity the ratio is high and stable, indicating that at maturity the cost to the system of maintaining large structure is small; photosynthesis per unit of biomass is low.

As in large natural systems, microecosystems become increasingly stratified and more heterogeneous during succession, and the importance of detritus as a source of energy and a location for many of the species increases. The algal bloom is replaced by large patches of detritus in which many of the bacteria and algae species are intimately associated. Presumably the ostracods, the top consumer in Beyers' and Cooke's ecosystems, graze on clumps of these organisms in the detritus.

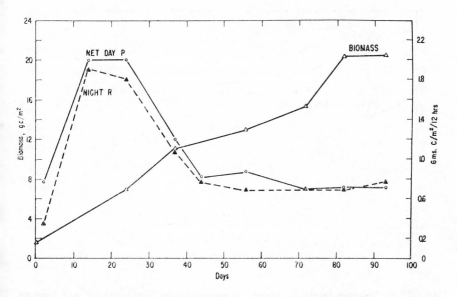

FIGURE 5. Pattern of net community photosynthesis, night respiration, and biomass changes during autotrophic succession of laboratory microecosystems. From Cooke (1967).

One of the most impressive changes in laboratory ecosystems during succession is the shift in color from bright green in early stages to a yellow-brown, often with small patches of green where ostracods are grazing, in mature stages (Cooke, 1967; Gorden, 1967; Taub, 1969). This gradual change in color accompanies the increase in stratification and the formation of detritus. Margalef (1965) has suggested that this shift in the ratio of pigments (carotenoids to chlorophylls) can be used as an index to the nutritional status of the autotrophic component of an ecosystem and thus to the age of the system. The Margalef pigment ratio is based on the premise that the commonly observed successional pattern is in part due to the gradual depletion of free plant nutrients and a closing of nutrient cycles as ecosystems age. Thus, as biomass accumulates, more and more of the available pool of nutrients is tied up and the rate of recycling becomes critical in determining the rate of ecosystem metabolism. That free nutrients are reduced in microecosystems is evidenced not only by the shift in color from green to yellow, a classic symptom of nutrient deficiency in algal culture and agriculture, but also to the gradual slowing of community metabolism.

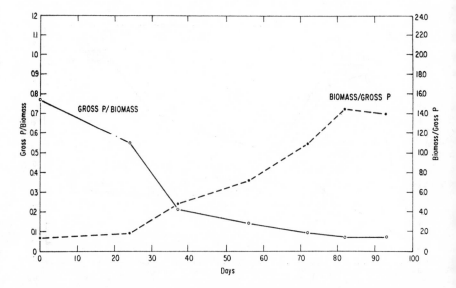

FIGURE 6. Changes in ratios between gross community photosynthesis and biomass (production efficiency) and biomass and gross community photosynthesis (maintenance efficiency) during autotrophic succession of laboratory microecosystems. From Cooke (1967).

The work of Whittaker (1961) offers further evidence to support these observations about succession in aquatic ecosystems and the availability of plant nutrients. Whittaker added radioactive phosphorus to aquaria-type microcosms and examined the rate of uptake and the recycling of the element during succession. As in large natural ecosystems (e.g., Hayes and Phillips, 1958; Rigler, 1964) phosphorus was taken up rapidly into plankton and then more slowly accumulated into other fractions of the ecosystem (Fig. 7). With low initial levels of phosphorus (oligotrophic aquaria), biomass developed toward a steady-state after 8 weeks of succession, with most of the added phosphorus in seston, side-wall and bottom algae, and sediment (Fig. 8). Very little was free in the water. In eutrophic aquaria steady-state was not reached during the term of the experiment, and, as pointed out by H. T. Odum and Hoskin (1957) for their enriched lotic ecosystem, succession was more rapid than in nutrient-poor systems. Whittaker also studied an artificial outdoor pond and observed a successional pattern of phosphorus transformations similar to that of the aquaria.

A shift in Margalef's pigment ratio may become a valuable index to the availability of plant nutrients and therefore to the age of the ecosystem if the pigment shift is caused by this progressive closing of nutrient cycles. Other biochemicals, such as lipids, may also be expected to increase as aquatic ecosystems age, just as in algal cultures which become nitrogen deficient. Thus there may be changes in the amounts and kinds of several biochemicals during succession so that biochemical diversity may be an important measure of succession in aquatic ecosystems. Some preliminary observations on the relationship between types of fatty acids and algal species composition and diversity in laboratory stream communities have been reported by McIntire, Tinsley and Lowry (1969). A quantitative study of the transformations of elements, particularly nitrogen and phosphorus, and the corresponding changes in plant pigments and other biochemicals during succession is needed.

In summary, the potential utility of microecosystems in the study of the ecosystem level of organization is supported by the fact that succession in the beaker-type microecosystem closely resembles both the long-term pattern of succession and many short-term seasonal cycles in large natural ecosystems. Thus, the trends in photosynthesis, respiration, and biomass during succession of a forest ecosystem and a microcosm are identical (E. P. Odum, 1969). As in the microcosms, forest succession leads to an ecosystem where rates of photosynthesis and respiration are nearly equal (Ovington, 1962). The early high rate of photosynthesis in an abandoned agricultural area gradually decreases (E. P. Odum, 1960), indicating that as succession towards a mature forest proceeds, the P/B ratio declines.

In a stable marine environment in which nutrients have been depleted the rate of metabolism is regulated by the rate of nutrient turnover (Ketchum et al., 1958). In some situations where nutrient cycles do not become

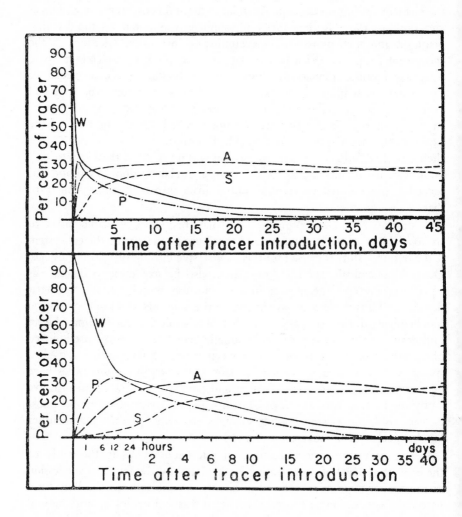

FIGURE 7. Experiment 1. p[32] distribution in oligotrophic aquaria (Columbia river water) in relation to time after tracer introduction, with time on linear scale above, square-root scale below. Vertical ticks above the base line for time indicates times of sample removal. Aquarium fractions are: water (W, solid line), algae attached to aquarium surface (A, longer dashes), plankton or seston (P, dot-and-dash), and sediment (S, shorter dashes). From Whittaker, (1961). Reproduced with permission of the author.

progressively closed but instead remain open and disturbed, the "bloom" stage of succession is maintained. Thus eutrophic lakes and the surrounding terrestrial area are like immature ecosystems. Microcosms dosed with large amounts of nitrogen and phosphorus, as in Abbott's (1967) study, have high productivity and algal blooms, just as in the immature stages of succession such as a eutrophic lake. Seasonal cycles of activity in ecosystems also exhibit the successional pattern. Thus, in the spring when amounts of nutrients, and temperature and light, are favorable, there follows a period of high productivity, biomass increase, and synthesis of

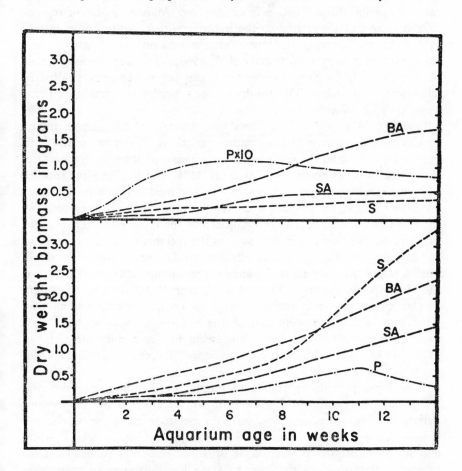

FIGURE 8. Development of aquarium communities, as expressed in dry-weight biomasses of community fractions, at low levels of phosphate fertilization above and high levels below. Aquarium fractions are: P—centrifuge plankton or seston (multiplied by 10 in the upper panel), SA—side wall algae, BA—bottom algae, and S—sediment, From Whittaker (1961). Reproduced with permission of the author.

chlorophyll, just as in the less mature stages of succession in microcosms. Later in the season, P/R and P/B ratios decline and the growth season ends. These seasonal cycles are imposed on the larger long-term pattern of ecological succession.

Regulation in Mature Microecosystems. The relationship between the rate of nutrient recycling and the metabolism of autotrophs at various stages of succession or the close coupling of photosynthesis and respiration, as demonstrated in Beyers' (1963a) photoperiod experiments, are examples of internal regulation in ecosystems. Another aspect of regulation in ecosystems is related to the hypothesis that mature multi-species ecosystems are far more resistant to external perturbations than are less mature or complex systems (Cooke et al., 1968). This hypothesis, difficult to examine in large natural ecosystems, can be conveniently studied in laboratory ecosystems. The results of some preliminary studies on this hypothesis are presented here.

Beyers (1962b) studied the metabolic response of his mature, multi-species microecosystem to temperature variations. Night respiration and net photosynthesis of the ecosystems were measured at room temperature, then after 24 hours acclimation, at about 33 °C, and finally, after another 24 hours, at temperatures of about 17°C. The total temperature variation was about 7 °C above and below the regular maintenance temperature for the ecosystems. The data were compared to the reported effects of temperature on *Daphnia* respiration and on the metabolism of a stream sewage community (Fig. 9). The results showed that respiration of the single species population was more dependent upon temperature than either of the multi-species systems. The balanced, complex laboratory ecosystem was less affected by temperature than the less complex sewage community. The drop in net photosynthesis of the microecosystems at high temperature was due either to an enhancement of day respiration or to an inhibition of photosynthesis by temperature. Increased temperature did not stimulate metabolism.

Beyers interpreted the results to mean that the more complex and integrated the system, the less likely it will be affected by temperature variations. Therefore, complex, mature ecosystems exhibit more thermal independence than do less complex or less mature ones. Beyers' study also suggests that the prediction of organism responses to habitat temperature variation from the results of studies of laboratory unispecific cultures may often not reflect the actual responses in multi-species situations.

While some investigators have reported data which supports Beyers' conclusions, the results of others do not. Whittaker (1961) observed that the differences in the rates of P[32] movement and rates of ecological succession in cool (10°C) and warm (25°C) laboratory ecosystems were much less than the 2-3 fold acceleration of rate expected with an increase of 10°C or more. Copeland and Dorris (1964) found that community

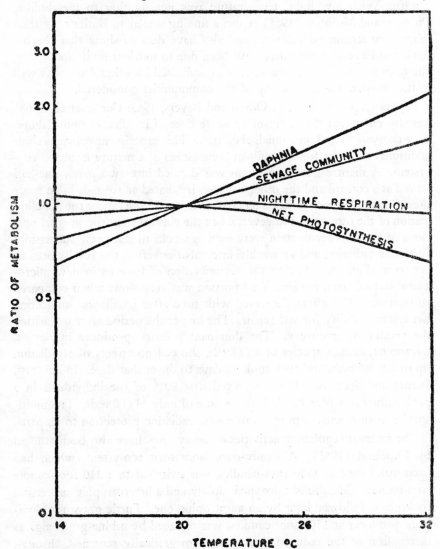

FIGURE 9. Effect of temperature upon the metabolism of a single organism, a sewage community, and the nighttime respiration and net photosynthesis of four fresh-water laboratory microecosystems. All measurements of metabolism are plotted as the ratio of metabolism at a given temperature to the metabolism of the same organism or system at 20°C. From Beyers (1962b). Reproduced with permission of the author.

respiration in oil refinery effluent holding ponds was about the same in both summer and winter. Butler (1964), on the other hand, reported that respiration of a pond-type microecosystem exhibited a 20% increase with a 12°C increase in temperature, when subjected to high light intensity. With low light, temperature was not limiting to metabolism. Phinney and McIntire (1965) report a finding similar to Butler's for their laboratory stream community, and also have data to show that the behavior of Beyers' system may have been due to nutrient limitation. Thus the response to temperature may be complicated by other factors as well as the comparative complexity of the communities considered.

Beyers (reported in E. P. Odum and Beyers, 1965; Cooke et al., 1968) has also compared the effect of an acute dose of irradiation upon laboratory ecosystems and on a unialgal culture. This experiment provides some additional evidence for the built-in homeostasis of a mature balanced ecosystem. A mature microecosystem was divided into two parts; one-half served as a control and the other half was irradiated at 10^6 rads in an acute dose. There were no visible effects from the irradiation except for the elimination of the ostracods. However, when the control and experimental portions of the microecosystem were used as stocks to start new successions just after radiation and at weekly intervals thereafter, the results became apparent (Fig. 10). At first the accumulation of biomass in new microcosms started from the irradiated portion was very slow, when compared to controls. The effect decreased with time after irradiation, indicating the system's ability for self-repair. The longer the period after irradiation the greater the recovery. The dominant primary producer in Beyers' microecosystem, a species of *Chlorella*, showed no effects of irradiation up to 2 x 10^6 rads, and then took 40 days to die at that dose. In contrast, Posner and Sparrow (1964) reported that 90% of the individuals in a pure culture of *Chlorella* died after a dose of only 23,000 rads. The multispecies mature ecosystem seems to confer radiation protection to its parts.

The internal regulatory activities of ecosystems have also been studied by Copeland (1965). A steady-state laboratory ecosystem, which had been maintained at 1500 foot-candles, was switched to a 230 foot-candle environment. Metabolism dropped rapidly and a heterotrophic succession took place, followed later by an autotrophic one. Turtle grass, the dominant producer at 1500 foot-candles, was replaced by a blue-green alga as metabolism of the reduced light ecosystem gradually returned, through the autotrophic succession, to the level of metabolism maintained by control turtle-grass ecosystems at 1500 foot-candles. Return to the stable, balanced original level of community metabolism was thus accomplished by a change in the dominant producer. Ecosystems apparently also have

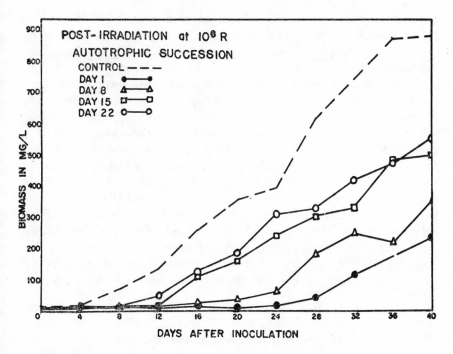

FIGURE 10. Course of biomass increase with time in an autotrophic succession in a laboratory microecosystem irradiated at 10^6 rad. Successions were initiated by inoculating samples of the irradiated mature microecosystems at 1, 8, 15, and 22 days after irradiation. Control curve is from non-irradiated microecosystem. From Cooke, Beyers, and Odum (1968).

a strong tendency to regulate to a balance between rates of photosynthesis and respiration.

Summary. Recent research on the flask or beaker-type of laboratory ecosystem has resulted in some important structural and functional data about the pattern of succession and the development of organization in ecosystems. Studies on these microcosms have also demonstrated certain regulatory properties of ecosystems which would have been very difficult to examine under experimental conditions in nature.

To summarize, the basic trend in ecological succession, as clearly demonstrated in microecosystems, is a shift from a structurally and metabolically unstable ecosystem to one of great homeostasis. The early high efficiency of production, an efficiency related to nutrient availability, shifts to a high maintenance efficiency at maturity, an efficiency apparently related to a tie-up of nutrients in biomass and a stable nutrient turnover rate. Further study will indicate whether measures of biochemical diversity, such as a plant pigment ratio, will serve as indices to the status of nutrients during succession in aquatic ecosystems. During succession, the population density and activity of the component species changes, and some species alter conditions in such a way as to favor the development of other species. Although not studied in microecosystems as yet, it is well known that some protists, including *Chlorella*, release inhibiting substances as well. Thus internal activities of these ecosystems alter conditions, to the extent permitted by energy input and other physical factors, to bring about an evolution of an ecosystem from a situation of early metabolic and structural instability to one of greater homeostasis. Mature ecosystems, as shown in the study of balanced laboratory microcosms, are apparently far more resistant to perturbations than are individual components of the system. The successional behavior of these laboratory ecosystems closely resembles that of large natural ecosystems, thus strengthening the contention that future study of microecosystems will prove to be of great value in understanding this level of biological organization.

Microcosms as Life Support Systems

The contrast between developmental and mature stages is very apparent in certain more practical approaches to ecosystems. In a recent symposium (Saunders, 1968), the progress in the development of a bioregenerative life support system for space travel was reviewed. For long-term space flight (hundreds or thousands of days), it is clear that a closed, completely regenerative life support system will be needed which will metabolize by-products of astronaut metabolism and will provide them with the necessary food, water, gases, and environmental diversity and protection in much the same way that these are provided in man's ecosystem on Earth. This life support system will properly be called a microecosystem, of which man is a component species.

One type of proposed microecosystem for space travel is to consist of only a few species (for example, *Chlorella*-man or *Hydrogenomonas*-man) plus a mass of supportive plumbing and back-up systems. Such an ecosystem would have the ecological characteristics of an immature system and would require a large amount of these energetically expensive external controls in order to prevent succession and to maintain long-term stability. Such an ecosystem, however, would have the advantage of high production efficiency (high P/B) and low biomass, thus reducing payload weight. Any failure of external homeostatic mechanisms or a change in the activity of the support organism would be fatal, since the system would then undergo ecological succession and might no longer support the astronaut. The probability of long-term reliability of a mechanical system with low biological diversity and the properties of an early successional stage is remote.

The alternative to the low diversity microcosm for life support in space is a multi-species mature ecosystem with its many built-in homeostatic mechanisms (H. T. Odum, 1963; Cooke et al., 1968; Cooke, 1971). As shown in studies of laboratory ecosystems, developmental stages go through a process of organization which results in an ecologically mature system which has long-term stability. Mature stages of ecosystems also seem to be more resistant to external perturbations, such as changes in temperature or an acute dose of radiation. Finally, mature stages, in contrast to developmental stages of microecosystems, are metabolically stable (Fig. 5), and the development of an AQ/RQ imbalance between astronauts and autotrophs would be very unlikely because in multi-species systems the RQ's of the several heterotroph species (including astronauts) would balance the AQ's of the several autotroph species, and a temporary imbalance with one species would be compensated by other species.

Several disadvantages of the multi-species microecosystem for life support in space have been shown by studies on laboratory ecosystems. The diurnal variation in ecosystem metabolism (Fig. 1) could lead to serious deleterious effects in any type of support system. Beyers' (1963b) data (Figs. 2 and 3) do indicate, however, that the amplitude of the daily variation is reduced with increasing diversity of the system. Also, the mature ecosystem has a low P/B ratio, and such a system would thus have to be very large (and heavy), when compared to the algae-man system, in order to support the astronauts.

The development of a reliable microecosystem for space will depend, in large part, upon studies of small laboratory microcosms and the application of the principles of ecology, particularly those dealing with ecological succession and the contrast between developmental and mature stages.

Laboratory Streams

Introduction

Laboratory streams are not ecosystems, in the sense that ecosystems have been defined by E. P. Odum (1963). They are, however, complex communities and some portion of the abiotic fraction of an ecosystem which have been allowed to reorganize and reassemble in the laboratory in an attempt to duplicate a large unit of nature. They are unnatural in the sense that, in some cases, there is no import nor influence of the terrestrial portion of the larger ecosystem. Also, the water is usually pre-filtered and large organisms are excluded.

Several different types of laboratory communities have been employed. H. T. Odum and Hoskin (1957) were perhaps the first to construct a model of a stream. Stokes (1960; also described in Kevern and Ball, 1965) built an aluminum trough, which was divided into six sections, each four feet long (Fig. 11). Depending upon the objectives of the experiments, there could be an 8 inch variation between sections, thus creating a series of riffles and pools. As in the Odum and Hoskin model, water was re-circulated. Lauff and Cummins (1964) also describe a stream of this type.

The most realistic duplication of a natural stream is the set of six replicate streams constructed by C. D. McIntire and his associates at Oregon State University (McIntire et al., 1964; Davis and Warren, 1965; Fig. 12). Water from a nearby stream is pumped into a wooden storage tank, then through a filter to the laboratory streams, and then out an effluent pipe. Water is not recirculated back to the mouth of the stream, and water chemistry, temperature, turbidity, organic load, and other factors vary with seasonal changes in the outside stream. The investigator can control such factors as light intensity, photoperiod, and current velocity. Small trays with rocks and rubble from the outside stream are placed in the troughs of the laboratory streams, and organisms in the incoming water seed the streams. Each tray can be used as a sample of a small area of the stream and can be removed to a photosynthesis-respiration chamber (Fig. 13) for brief measurements of community metabolism under carefully controlled conditions. McIntire's model stream closely resembles a section of a natural stream, in the sense that rates of photosynthesis, their efficiency of light fixation, the amount of chlorophyll—a per square meter, their species composition, and the light intensity, photoperiod, and current velocities are very similar to values which have been regularly reported for a variety of natural streams (McIntire and Phinney, 1965). The replicability of the six streams, at least with respect to rates of community metabolism, was clearly demonstrated (McIntire, et al., 1964), and the pattern of total community energy flow closely duplicated that reported for natural streams (Davis and Warren, 1965).

The following discussion of laboratory streams will be confined to research on two important problems in lotic ecology: methods of metabolism measurement, and stream succession.

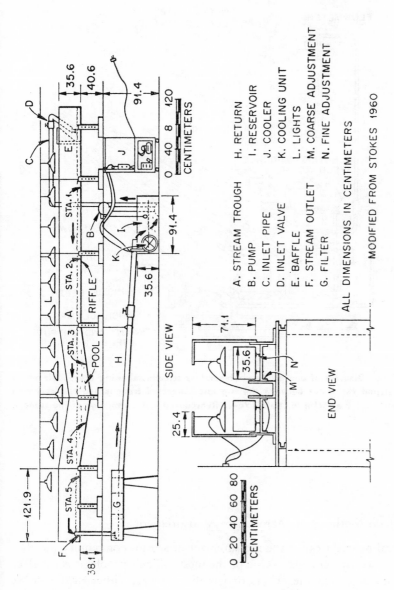

A. STREAM TROUGH   H. RETURN
B. PUMP   I. RESERVOIR
C. INLET PIPE   J. COOLER
D. INLET VALVE   K. COOLING UNIT
E. BAFFLE   L. LIGHTS
F. STREAM OUTLET   M. COARSE ADJUSTMENT
G. FILTER   N. FINE ADJUSTMENT

ALL DIMENSIONS IN CENTIMETERS

MODIFIED FROM STOKES 1960

FIGURE 11. Diagram of artificial streams. From Kevern and Ball (1965). Reproduced with permission of the authors.

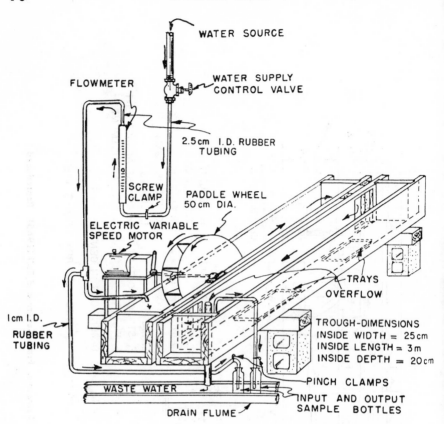

FIGURE 12. Diagram of one of the six laboratory streams, showing the paddle wheel for circulating the water between the two interconnected troughs and the exchange water system. From McIntire et al. (1964). Reproduced with permission of the authors.

Studies on Methods of Metabolism Measurement

Several procedures for the measurement of stream community metabolism are used by ecologists. One of the most difficult problems is to select a method in which the effects of gas diffusion are either negligible or unimportant, or can be accurately corrected for. The utility of laboratory streams to examine this problem was clearly demonstrated in a study by Kevern and Ball (1965), who compared the effects of five factors,

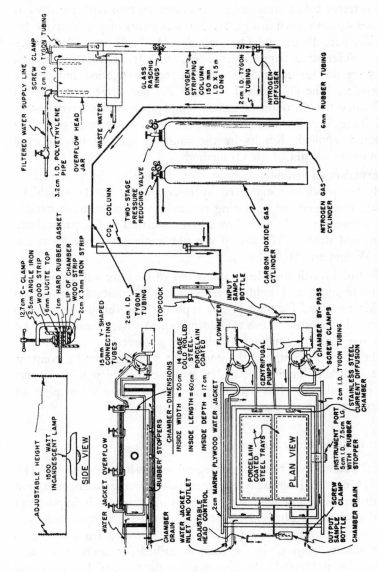

FIGURE 13. Diagram of the photosynthesis-respiration chamber, showing the chamber with its circulating and exchange water systems, the water jacket for temperature control, the nutrient and gas concentration control system, and the light source. From McIntire et al. (1964). Reproduced with permission of the authors.

temperature, current velocity, light intensity, photoperiod, and addition of chelates, on the results obtained by measuring community metabolism by the dissolved oxygen, the pH, and the substrate (harvest) methods. In the experiment, conditions were kept constant and the effect of a single variable was tested by creating differences between two replicate streams built like the stream shown in Fig. 11. Measurements of productivity by the pH (Verduin, 1960) and O_2 methods (H. T. Odum, 1956) were significantly different in temperature, light, and current velocity studies, but there were no significant differences in the photoperiod and chelate addition studies. The O_2 and pH methods did not agree when factors which influence the diffusion of gases (temperature and current) were compared, indicating that diffusion is a major source of error, even when a correction is applied. Results by the substrate method were always lower than either the pH or the O_2 methods.

These experiments strongly suggest that the use of the pH method, particularly the procedure suggested by Beyers et al. (1963), should be more widely employed for measuring community metabolism in streams since CO_2 diffusion may be very low when compared to O_2, particularly in well-buffered waters. The use of the pH method for streams is described by Wright and Mills (1967).

A number of investigators have attempted to measure stream productivity from pigment and light data, assuming a constant assimilation number. McIntire and Phinney (1965) reported that this number (ratio gross community photosynthesis to chlorophyll-a) varied from 0.2 to 1.4 milligrams oxygen/hour/milligram chlorophyll-a over a year's interval in their laboratory streams, and the conditions which favored a constant ratio were constant illumination and a stable community with respect to species composition. Their data suggest caution in the use of pigment measurements to estimate productivity in streams.

Succession in Laboratory Streams

In natural streams, two types of ecological succession are observed (H. T. Odum, 1956): 1) longitudinal upstream-downstream succession, and 2) the short-term response of a stream section to variations in import and to seasonal changes in physical factors such as light, temperature, and current velocity. Streams have a gradation of maturity from source to mouth but stream "climax," in the sense described for beaker-type ecosystems, must be interpreted in relation to a broader terrestrial-aquatic type of climax ecosystem (Margalef, 1960). A terrestrial-aquatic model stream has not yet been constructed and present laboratory streams have insufficient length to exhibit an upstream-downstream effect.

Studies of ecological succession in laboratory streams have been concentrated almost exclusively on short-term succession. Seasonal changes in import and local physico-chemical conditions produce a pronounced response in the structure and metabolism of any given stream section, and if these conditions remain fairly constant, the stream will come to a quasi-steady state relative to the conditions. This is short-term ecological succession in streams. One of the difficulties in the analysis of this type of succession in nature is to determine which factors are most important in determining a particular level of structure and function of a stream section. Laboratory streams, where all factors but one or a combination may be held constant, have been very useful in providing data about this problem.

As indicated earlier, McIntire's stream is constructed so that water from a natural stream flows through it. The incoming water varies seasonally in quality, and in the laboratory various physical regimes may be imposed. Two different approaches to the study of seasonal succession in his streams have been made. In one set (McIntire and Phinney, 1965) two streams, a light-adapted and a shape-adapted, were studied. Length of the daily photoperiod was altered every two weeks to correspond to seasonal changes, and seasonal variations in the structure and metabolism of these two streams were then examined. In the other study (McIntire, 1968a) six streams were set-up, three at 150 foot-candles and three at 700 foot-candles. Three current velocities, 0, 14, and 35 cm./second were established within each light regime, and, as usual, water quality varied with seasonal changes in the outside source stream. Depending on these seasonal variables and the differences in current and light, certain algal species were most abundant in each of the six streams, and the streams had characteristic levels of biomass, metabolism, and export.

An analysis of the response of benthic algal species to the six current and light regimes was made. Current velocity and low light had a statistically significant positive effect on most of the diatom species (79 of the 89 species in the streams were diatoms), but a few diatoms were most successful at high light and low current. McIntire's (1968a) data indicate that at any given season there is an algal community with a characteristic species composition for that season. An example, the seasonal responses of populations of three algal species *Nitzschia linearis* (Bacillario phyta), *Phormidium retzii* (Cyanophyta), and *Tribonema minor* (Chrysophyta) in the six different streams, is shown in Fig. 14. The data of Fig. 14 illustrate the value of controlled laboratory communities in autecological analyses. The algal populations grow under conditions much like that found in natural streams, in contrast to the artificiality of unialgal cultures, and the investigator can hold all but one or two factors constant.

74   G. DENNIS COOKE

A month or more was required for these communities to reach a quasi-steady state of biomass (Fig. 15), and this level fluctuated considerably, depending upon season or experimental conditions. During periods of seasonal change, breakdown and recolonization were very rapid. The rate of biomass development was greater in light-adapted than in shade-adapted, and also greater in streams with the fastest current (McIntire and Phinney, 1965; McIntire, 1966a, 1968,b). The greater rate of biomass accumulation in the faster currents was attributed to a more efficient use of the substrate and to enhancement of metabolism by current.

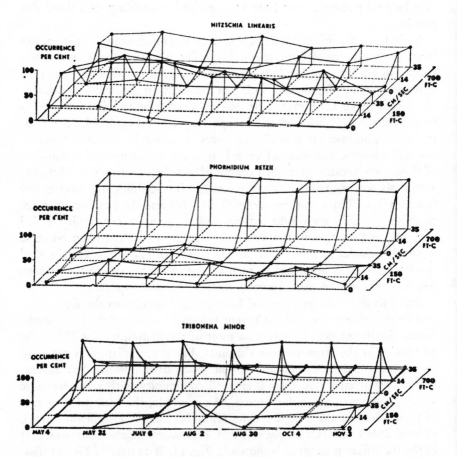

FIGURE 14. Observed occurrence of *Nitzschia linearis, Phormidium retzii,* and *Tribonema minor* in the six laboratory streams on the seven sampling dates during the experiment. Occurrence is expressed as the percentage of microscope fields in which the organisms were observed. From McIntire (1968a). Reproduced with permission of the author.

Changes in community metabolism were shown to be related to the previous history of illumination of the community, and to water temperature, velocity, and amount of nutrients. Stream sections which develop and are exposed to higher light intensity had a higher rate of photosynthesis (Fig. 16). The enhancement of the photosynthetic rate with an increase in light was linear up to about 1000 foot-candles in communities developed at 500 foot-candles (Phinney and McIntire, 1965; McIntire and Phinney, 1965; McIntire, 1968a), and shade-adapted communities reach light saturation at values only slightly less than light-adapted communities. McIntire (1968b) observed that saturation for shade-adapted cells in unialgal cultures is usually much less than saturation for light-adapted cells of the same species, and he interprets these observations to

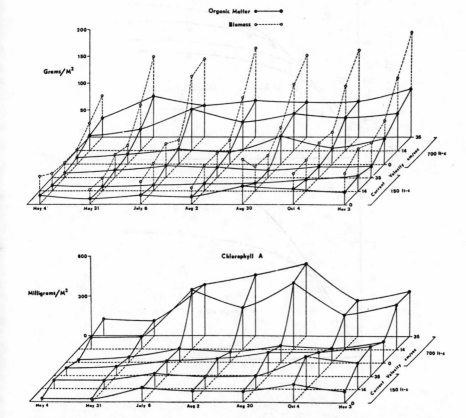

FIGURE 15. The biomass (g/M²), organic matter (g/M²), and chlorophyll-a (mg/M²) in the six laboratory streams on the seven sampling dates during the experiment. From McIntire (1968A). Reproduced with permission of the author.

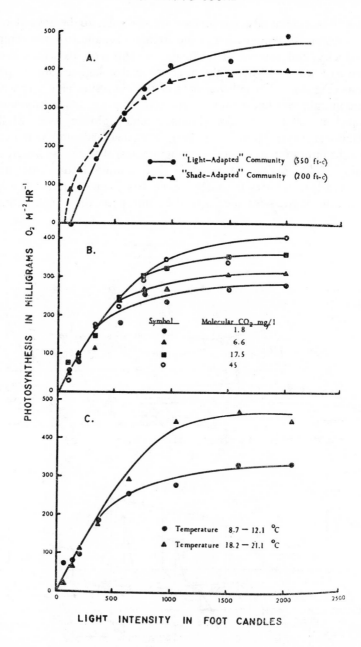

FIGURE 16. The relationship between light intensity and rate of photosynthesis for periphyton communities developed in the laboratory is influenced by: A. light adaptation, B. concentration of carbon dioxide, and C. temperature. From McIntire (1968b). Reproduced with permission of the author.

mean that in complex communities light absorption is more heterogeneous due to their species diversity. One could further state that the response of a stream community to physical factors is modified by the community itself and cannot be predicted from the responses of individual species in laboratory cultures.

In contrast to the hypothesis of Beyers (1962b) that the metabolism of a complex, interacting assemblage of organisms will exhibit a considerable amount of temperature independence, an increase in the temperature of experimental streams usually leads to an increase in stream metabolism. The relationship between community respiration and temperature between 6 and 17°C was exponential. The Q_{10} was usually about 2.0, with the lowest Q_{10} values reported for high temperature ranges (Phinney and McIntire, 1965; McIntire, 1966b). At lower temperature ranges (3-13°C), slow-current communities were more sensitive to temperature changes; at the higher temperature ranges (13-23°C) the reverse was true. When light intensity was near 2000 foot-candles, an increase from 12 to 20°C resulted in an increase in gross community photosynthesis, but at light levels below 1000 foot-candles (Fig. 16) there was no significant increase in photosynthesis, indicating that communities exposed to this intensity are light-limited (Phinney and McIntire, 1965; McIntire, 1968a). Butler (1964) reported a similar observation for pond-type microcosms. These streams have considerably more species diversity than Beyers' (1962b) beaker-type systems (McIntire, personal communication) and thus at least as much internal homeostasis. It is possible that Beyers' systems were light-limited.

The rate of current velocity is frequently believed to be a very important factor in the rate of community metabolism because rapid water flow apparently produces a steeper diffusion gradient to absorbing cells and thus facilitates exchange between cells and water (Whitford and Schumacher, 1961). McIntire (1966a,b; 1968a) has shown that the metabolism of stream communities is enhanced by high current velocity as well as temperature and light. His data indicate that communities that develop in rapid current exhibit a greater metabolic response to changes in current velocity than do communities that develop in slow current.

In related studies, McIntire and Phinney (1965) and McIntire (1968b) reported that stream communities are very susceptible to CO_2 exhaustion. In light-adapted communities at high illumination, an addition of molecular CO_2 resulted in an increase in photosynthesis (Fig. 16). Shade-adapted streams do not respond to addition of CO_2, apparently because they are light-limited. When streams are CO_2-limited, Q_{10} ranges from

1.0 to 1.5. Thus Beyers' (1962b) temperature-independent microcosm could also be CO_2 (or other nutrient)-limited in addition to a possible light limitation because recycling in those systems may not be rapid at maturity and they have very little current.

The level of biomass at any given time in the season was related not only to those factors that regulate community metabolism, but also to the rate of export from the stream section. In laboratory streams, peaks of export were often found just after a biomass development peak. As precipitation increased in the fall the silt load entering the laboratory stream increased and apparently had a scouring effect. Communities which developed at low light had many diatoms and were not as vulnerable to export, while light-adapted communities had many filamentous algae which were easily dislodged. Both slow-current and fast-current communities often had nearly identical biomass levels, even though export from the fast-current community was much higher. Fast-current streams had a higher rate of net community photosynthesis.

The amount of chlorophyll-a per square meter of stream community and the ratio of carotenoids to chlorophyll-a have been shown to be reliable indices to the conditions to which the stream has been exposed (McIntire and Phinney, 1965; McIntire, 1968a). The greatest concentrations of chlorophyll-a were found in communities that developed at high light and rapid current (Figure 15). The Margalef pigment index (Margalef, 1965) was highest in streams developed at high light and no current (pond-like), indicating that nutrient depletion can be expected in stream communities exposed to near stagnant conditions. During the summer, the index remained fairly stable in all streams, but increased during the onset of fall when temperature and nutrient supply were decreased. McIntire's data indicate this index to be a rather sensitive one to changes in the nutrition of algal communities in streams.

To summarize, the rate at which a community is established through ecological succession in any particular section of a stream, as shown by studies of laboratory streams, is related to current velocity, light intensity, and previous illumination history, and such factors as silt load. The rate of photosynthesis during a season is often light-limited and will be greater in sections with high current and high export. Temperature is usually not a factor on a day-to-day basis because it remains rather steady over a season, except in slow-current areas, and is effective only at high light intensities (above 1000 foot-candles). The type of community that develops, in terms of species, is related to current and light. Some communities have a particular morphology (microstratification) that may inhibit the diffusion of gases and nutrients or may be self-shading. It appears that the Margalef pigment index may be very sensitive to changes in the plant nutrient level in streams. McIntire has recently summarized some of his studies with laboratory streams (McIntire, 1969).

The study of laboratory streams, where the effects of single factors on community activity may be analyzed, and where the rate and type of succession over an entire year under various conditions of illumination and current speed can be examined, has provided some valuable data which will be very useful in interpreting the overall pattern of longitudinal succession and organization of a natural stream. The development of a terrestrial-aquatic laboratory stream, with upstream-downstream succession, would be very valuable for future investigations.

Conclusions

The preceding discussions have provided considerable evidence to show that laboratory microecosystems and complex communities have the properties of similar larger natural systems. This has been demonstrated in laboratory studies of diurnal community metabolism, ecological succession, and community regulation, as well as in experimental investigations on the ecology of one or more small groups of species.

The use of "laboratory strains" of organisms is well known in biology. These have been developed as a means of testing hypotheses and for suggesting theories through experimentation on a controlled and defined but complex unit of biological organization. Thus *Drosophila*, *Neurospora*, *Escherichia coli*, and *Tetrahymena* have been thoroughly studied by geneticists and cell biologists. Similarly, animal physiologists have used the white rat, embryologists the frog, and population ecologists the flour beetle and *Daphnia*. The laboratory microcosm may well become an ecological "white rat" for the study of the ecosystem level of biological organization. As illustrated by laboratory streams, and by the recent use of experimental laboratory systems to examine the effect of simulated tidal cycles on marine benthic diatom communities (McIntire, 1969; McIntire and Wulff, 1969), autecological studies, as well as investigations about more complex levels of organization, may be conducted in the laboratory without resorting to the artificiality of unispecific cultures.

In addition to the study of the basic principles of ecology, the laboratory community or ecosystem may also become a very valuable diagnostic tool. For example, many new organic compounds are developed each year in this country, and most of these new compounds are released to the environment before we understand all of the possible effects of them. The usual procedure, if any testing is done at all, is to do toxicity tests on one or more vertebrate species (game species or species with many genes in common with man). Unfortunately, such tests are relatively meaningless since pollutants such as pesticides are applied, in practice, to the community and ecosystem levels of organization and not to these organism or population levels. Laboratory ecosystems will make excellent devices for the examination of the short and long-range effects of potential pollutants on the actual level of biological organization to which they are applied. Such studies can be carried out before the potential pollutants are released. Two recent examples (Hueck and Adena, 1968; Cairns, 1969) illustrate the effect of some common pollutants on the activity of laboratory ecosystems and communities.

A logical extension of the laboratory microcosm method is to construct miniature, replicable outdoor ecosystems and communities. Thus, Whitworth and Lane (1969) used a series of wading pools as controlled ecosystems in order to assess the long-term effect of a group of toxicants on ponds. The use of polyethylene bags or other containers to isolate a portion of the plankton community for experimental purposes is well known (e.g., Antia et al., 1963; Strickland et al., 1969). As H. T. Odum et al. (1963) pointed out, the construction of outdoor systems is actually an experiment in ecological engineering which may result in some new systems of considerable economic value to man.

Finally, the microcosm approach is very useful in teaching, since the student is continuously presented with a holistic view of ecology and at the same time is provided with a system amenable to laboratory and field experimentation (Bovbjerg and Glynn, 1960).

In conclusion, it should be pointed out that there are several very interesting questions in ecology which may be answered through the microcosm method. For example, what is the minimum ecosystem (in terms of area and diversity) for a man or a group of men (space flight)? What is the relationship between species and pattern diversity and the structural and metabolic stability of an ecosystem?

Acknowledgments

Preparation of this manuscript was supported, in part, by a Kent State University Summer Research Fellowship. I thank Drs. C. David McIntire and R. W. Gorden for their constructive criticisms of the manuscript. To Drs. R. J. Beyers, R. W. Gorden, N. R. Kevern, C. D. McIntire, and R. H. Whittaker, I owe a particular debt of gratitude for their permission to reproduce some of their figures for use in this paper.

Literature Cited

Abbott, W. 1966. Microcosm studies on estuarine waters. I. The replicability of microcosms. J. Wat. Poll. Cont. Fed., 38: 258-270.

——————— 1967. Microcosm studies on estuarine waters. II. The effects of single doses of nitrate and phosphate. J. Wat. Poll. Cont. Fed., 39: 113-122.

Antia, N. J., C. D. McAllister, T. R Parsons, and J. D. H. Strickland. 1963. Further measurements of primary production using a large-volume plastic sphere. Limnol. Oceanogr., 8: 166-183.

Beyers, R. J. 1962a. The metabolism of twelve aquatic laboratory microecosystems. Ph.D. dissertation. The University of Texas.

——————— 1962b. Relationship between temperature and the metabolism of experimental ecosystems. Science, 136: 980-982.

——————— 1963a. The metabolism of twelve laboratory microecosystems. Ecol. Monogr., 33: 281-306.

——————— 1963b. A characteristic diurnal metabolic pattern in balanced microcosms. Publ. Inst. Mar. Sci., Texas, 9: 19-27.

——————— 1963c. Balanced aquatic microcosms——their implications for space travel. Amer. Biol. Teach., 25: 422-429.

——————— 1965. The pattern of photosynthesis and respiration in laboratory microecosystems. Mem. Ist. Ital. Idrobiol., 18 suppl.: 61-74.

———————, J. L. Larimer, H. T. Odum, R. B. Parker, and N. E. Armstrong. 1963. Directions for the determination of changes in carbon dioxide concentration from changes in pH. Publ. Inst. Mar. Sci., Texas, 9: 454-489.

Bovbjerg, R. V. and P. W. Glynn. 1960. A class exercise on a marine microcosm. Ecol., 41: 229-232.

Butler, J. L. 1964. Interaction of effects by environmental factors on primary productivity in ponds and microecosystems. Ph.D. dissertation, Oklahoma State University.

Cairns, J., Jr. 1969. Rate of species diversity restoration following stress in freshwater protozoan communities. The Univ. Kansas Sci. Bull., XLVIII: 209-224.

Coler, R. A. and H. B. Gunner, 1970. Laboratory enclosure of an ecosystem response to a sustained stress. Appl. Microbiol., 19: 1009-1012.

Cooke, G. D. 1967. The pattern of autotrophic succession in laboratory microcosms. BioScience, 17: 717-721.

——————— 1971. Space Ecology: The minimum ecosystem for man. In: E. P. Odum. Fundamentals of ecology, 3rd edition. W. B. Saunders Co., Philadelphia.

———————, R. J. Beyers, and E. P. Odum. 1968. The case for the multi-species ecological system, with special reference to succession and stability. In: J. F. Sauners (Ed.) Bioregenerative systems. NASA SP-165: 129-139.

Copeland, B. J. 1965. Evidence for regulation of community metabolism in a marine ecosystem. Ecol., 46: 563-564.

———————— and T. C. Dorris. 1964. Community metabolism in ecosystems receiving oil refinery effluents. Limnol. Oceanogr., 9: 431-447.

Davis, G. E. and C. E. Warren. 1965. Trophic relations of a sculpin in laboratory stream communities. J. Wildl. Management, 24: 846-871.

Goering, J. J., R. C. Dugdale, and D. W. Menzel. 1964. Cyclic diurnal variations in the uptake of ammonia and nitrate by photosynthetic organisms in the Sargasso Sea. Limnol. Oceanogr., 9: 448-451.

Gorden, R. W. 1967. Heterotrophic bacteria and succession in a simple laboratory aquatic microcosm. Ph.D. dissertation, The University of Georgia.

————————, R. J. Beyers, E. P. Odum, and R. G. Eagon. 1969. Studies of a simple laboratory microecosystem: Bacterial activities in a heterotrophic succession. Ecol., 50:86-100.

Hayes, F. R. and J. E. Phillips. 1958. Lake water and sediment. IV. Radiophosphorus equilibrium with mud, plants, and bacteria under oxidized and reduced conditions. Limnol. Oceanogr., 3: 459-475.

Hueck, H. J. and D. M. M. Adena. 1968. Toxicological investigations in an artificial ecosystem. A progress report on copper toxicity towards algae and *Daphnia*. Helgolander Wiss. Meersunters, 17: 188-199.

Jackson, D. R. and J. McFadden. 1954. Phytoplankton photosynthesis in Sanctuary Lake, Pymatuning Reservoir. Ecol., 35: 1-4.

Ketchum, B. H., J. H. Ryther, C. S. Yentsch, and N. Corwin. 1958. Productivity in relation to nutrients. Rapp. Proc.-Verb. Cons. Int. Explor. Mer., 144: 132-140.

Kevern, N. R. and R. C. Ball. 1965. Primary productivity and energy relationships in artificial streams. Limnol. Oceanogr., 10:74-87.

Kurihara, Y. 1960. Biological analysis of the structure of microcosm, with special reference to the relations among biotic and abiotic factors. Sci. Rep. Tohoku Univ. Ser. IV (Biol.), 26: 269-296.

Lauff, G. H. and K. W. Cummings. 1964. A model stream for studies in lotic ecology, Ecol., 45: 188-190.

Margalef, R. 1958. Temporal succession and spatial heterogeneity in phytoplankton. In: A. A. Buzzatti-Traverso (Ed.). Perspectives in Marine Biology. The University of California Press, Berkeley. pp. 323-349.

———————— 1960. Ideas for a synthetic approach to the ecology of running water. Int. Rev. ges. Hydrobiol., 45: 133-153.

———————— 1963a. On certain unifying principles in ecology. Amer. Nat., 97: 357-374.

———————— 1963b. Succession of marine populations. Adv. Frontiers of Plant Sci., 2: 137-188 (India).

——————— 1965. Ecological correlations and the relationship between primary productivity and community structure. Mem. Ist. Ital. Idrobiol., 18 suppl.: 355-364.

——————— 1968. Perspectives in Ecological Theory. The University of Chicago Press. Chicago. viii+ 111 pp.

McConnell, W. J. 1962. Productivity relations in carboy microcosms. Limnol. Oceanogr., 7: 335-343.

——————— 1965. Relationship of herbivore growth to rate of gross photosynthesis in microcosms. Limnol. Oceanogr., 10: 539-543.

McIntire, C. D. 1966a. Some effects of current velocity on periphyton communities in laboratory streams. Hydrobiol., 27: 559-570.

——————— 1966b. Some factors affecting respiration of periphyton communities in lotic environments. Ecol., 47: 918-930.

——————— 1968a. Structural characteristics of benthic algal communities in laboratory streams. Ecol., 49: 520-537.

——————— 1968b. Physiological-ecological studies of benthic algae in laboratory streams. J. Wat. Poll. Cont. Fed., 40: 1940-1952.

——————— 1969. A laboratory approach to the study of the physiological ecology of benthic algal communities, In: Proceedings of the eutrophication-biostimulation assessment workshop, Berkeley, Calif. E. J. Middlebrooks, T. E. Maloney, C. F. Powers and L. M. Kaack (Editors). pp. 146-157.

———————, R. L. Garrison, H. K. Phinney, and C. E. Warren. 1964. Primary production in laboratory streams. Limnol. Oceanogr., 9: 92-102.

——————— and H. K. Phinney. 1965. Laboratory studies of periphyton production and community metabolism in lotic environments. Ecol. Mongr., 35: 237-258.

———————, I. J. Tinsley, and R. R. Lowry. 1969. Fatty acids in lotic periphyton: another measure of community structure. J. Phycol., 5: 26-32.

———————and B. L. Wulff. 1969. A laboratory method for the study of marine benthic diatoms. Limnol. Oceanogr., 14: 667-678.

Nixon, S. W. 1969. A synthetic microcosm. Limnol. Oceanogr., 14: 142-145.

Odum, H. T. 1956. Primary production in flowing water. Limnol. Oceanogr., 1: 102-117.

——————— 1963. Limits of remote ecosystems containing man. Amer. Biol. Teach., 25: 429-443.

——————— and C. M. Hoskin. 1957. Metabolism of a laboratory stream microcosm. Publ. Inst. Mar. Sci., Texas, 4: 115-133.

——————, R. J. Beyers, and N. E. Armstrong. 1963. Consequences of small storage capacity in nannoplankton pertinent to measurement of primary production in tropical waters. J. Mar. Res., 21: 191-198.

——————, W. L. Siler, R. J. Beyers, and N. E. Armstrong. 1963. Experiments with engineering of marine ecosystems. Publ. Instit. Mar. Sci., Texas, 9: 373-403.

Odum, E. P. 1960. Organic production and turnover in old field succession. Ecol., 41: 34-49.

—————— 1963. Ecology. Holt, Rinehart, and Winston. New York. vii+ 152 pp.

—————— 1969. The strategy of ecosystem development. Science, 164: 262-270.

—————— and R. J. Beyers. 1965. Biodynamics of microecosystems. Report to NASA, 1 August, 1965. 6 pp.

Ovington, J. D. 1962. Quantitative ecology and the woodland ecosystem concept. Adv. Ecol. Res., 1: 103-192.

Owens, M. 1965. Some factors involved in the use of dissolved-oxygen distributions in streams to determine productivity. Mem. Ist Ital. Idrobiol., 18 suppl.: 209-224.

Phinney, H. K. and D. D. McIntire. 1965. Effect of temperature on metabolism of periphyton communities developed in laboratory streams. Limnol. Oceanogr., 10: 341-344.

Posner, H. B. and A. H. Sparrow. 1964. Survival of *Chlorella* and *Chlamydomonas* after acute and chronic gamma radiation. Rad. Bot., 4: 253-257.

Rigler, F. H. 1964. The phosphorus fractions and the turnover time of inorganic phosphorus in different types of lakes. Limnol. Oceanogr., 9: 511-518.

Saunders, J. F. (ed.) 1968. Bioregenerative Systems. NASA Special Publication 165. viii+ 153 pp. U.S. Gov. Printing Office, Washington, D. C.

Shimada, B. M. 1958. Diurnal fluctuation in photosynthetic rate and chlorophyll "A" content of phytoplankton from eastern Pacific waters. Limnol. Oceanogr., 3: 336-339.

Stokes, R. M. 1960. The effects of limiting concentrations of nitrogen on primary production in an artificial stream. M. S. Thesis. Michigan State University.

Strickland, J. D. H., O. Holm-Hansen, R. W. Eppley, and R. J. Linn. 1969. The use of a deep tank in plankton ecology. I. Studies of the growth and composition of phytoplankton crops at low nutrient levels. Limnol. Oceanogr., 14: 23-34.

Taub, F. B. 1969. A biological model of a freshwater community: A gnotobiotic ecosystem. Limnol. Oceanogr., 14: 136-142.

—————— and A. M. Dollar. 1964. A *Chlorella-Daphnia* food-chain study: The design of a compatible chemically-defined culture medium. Limnol. Oceanogr., 9: 61-74.

Verduin, J. 1960. Differential titration with strong acids or bases vs. CO_2 water for productivity studies. Limnol. Oceanogr., 5: 228.

Whittaker, R. H. 1961. Experiments with radiophosphorus tracer in aquarium microcosms. Ecol. Monogr., 31: 157-188.

Whitworth, W. R. and T. H. Lane. 1969. Effects of toxicants on community metabolism in pools. Limnol. Oceanogr., 14: 53-58.

Wright, J. C. and I. K. Mills, 1967. Productivity studies on the Madison river, Yellowstone National Park. Limnol. Oceanogr., 12: 568-577.

George W. Salt

The Role of Laboratory Experimentation in Ecological Research

Abstract

Contemporary ecological research seeks to produce principles characterized by realism, precision, generality, and the capacity to predict future events in natural ecosystems. Various research techniques may be employed for this purpose. It is suggested here that one fruitful line of procedure is as follows: Hypotheses are generated from observations in the field. These are then tested and quantified by laboratory experimentation. Experimental results are next used to construct a computer model from which predictions of events in nature are generated. These predictions are verified in the field. If necessary, the initial hypotheses are modified and the entire circuit repeated.

It is further suggested that if an ecological phenomenon known in higher animals can be discerned in micro-organisms, it is likely to be a general one. Micro-organisms have the additional advantage of being particularly appropriate for laboratory experimentation.

Introduction

There exists today a pressing need for predictive ecological theory. Many of our most urgent public questions are ecological ones requiring decisions on resource utilization, correction of pollution of the air and water, and control of noxious organisms, either agricultural pests or agents of human disease. Further, there is good reason to believe that ecological theory may be applied in some degree to human affairs in the solution of urban and economic problems.

Discussion

We can make reasonably confident statements about certain natural phenomena. This is not the place to list these generalizations in detail. In brief, however, we are aware of the probable changes in species diversity, biomass, stability, and productivity with time in both terrestrial and aquatic communities (Odum, 1969; Margalef, 1968; Hutchinson, 1967). We appreciate the role of physical diversity in increasing species diversity (MacArthur, 1965) and stability in predator-prey interactions (Huffaker, 1958; Huffaker, Shea and Herman, 1963). If two species both compete for the same limited resource, we can make an informed guess as to the probable winner of the contest (Hairston and Kellermann, 1965; Miller, 1967; Morse, 1967; Evans, 1958). We are beginning to comprehend the interaction between self-regulation and environmental regulation in the control of population density (Sluss, 1967; Munro, 1967).

At first sight this is a rather impressive list of statements. It does represent an improvement over the situation twenty years ago when ecology was largely based on succession, Liebig's law of the minimum, and Gause's law of competitive exclusion. However, many of today's generalizations are only qualitative in nature. They lack precision and, to some degree, are too largely descriptive. Further, there are vast areas of ecological interactions that are not well comprehended.

It is generally accepted that any model, theory, or mathematical description of an ecological phenomenon should possess three qualities. These are generality, precision, and realism (Holling, 1963, 1966a, 1966b). A fourth quality, which I consider essential, is the capacity for prediction. Unless a generalization permits the judging of the outcome of future ecological interactions with a usable degree of accuracy it is merely a passive description. Ecologists use different approaches to achieve these goals. I should like here to examine these various strategies and to identify the role of laboratory experimentation in them.

One approach starts with the premise that the simultaneous attainment of equal degrees of precision, generality, and realism is an impossibility. One must be sacrificed to maximize the other. A recent book by Levins (1968) presents this point of view in detailed form. Here the paramount goal is generality, for the attainment of which both precision and realism are abandoned. Study is conducted on models composed of "sufficient parameters" containing the relevant aspects of reality for the purpose of the model. Manipulation of these models by mathematical means leads to the formulation of a "robust theorem." Such a theorem or hypothesis is a high level abstraction in which ecological processes are related in terms of their inequalities, relative rates, or some other relationship. The hope is that one can progress downward through lesser degrees of abstraction from a theorem by the addition of more and more realistic information to the point where a concrete statement can be made about events in nature.

One may raise several questions about this approach. It remains to be demonstrated that specific application of such a high level abstraction can be accomplished. It further remains to be shown that the time and effort required to derive the abstraction in any way expedites the formulation of conclusions at the level of real animals and plants. It may well be that one could have reached the same statement of relationships by simply studying the process in detail without resorting to the formulation of the "robust theorem."

A second procedure is to state ecological relationships in the language of established mathematics. Hutchinson described the niches of competing species and the results of their interactions in terms of set theory (Hutchinson, 1957). Margalef (1968) used information theory to describe the composition and change in plankton. MacArthur (1955) made predictions on community stability based on the assumption that feeding relationships could be considered Markov processes. Preston (1962) described the species composition of faunas of related forms in terms of a canonical relationship. There are many other examples. The attraction of this method is that if it can be demonstrated that organisms do, in fact, fit such mathematical formulations, one can then manipulate the mathematics according to accepted rules to derive statements of relationship that were not previously apparent (Garfinkel, 1967). In the majority of instances the results of this method have been descriptions of processes which were more explicit and precise than was attainable before. The degree of generality and realism that such formulations contain has not been rigorously tested. However, in the few cases in which predictions have been made from such statements, they have not proven accurate, which suggests that the degree of generality or realism is not high (Hairston, et al., 1968).

Holling (1963) established the use of component analysis in the study of ecological processes. The method consists of subdividing an ecological phenomenon into its constituent parts. Experiments are then performed which lead to the formulation of a mathematical description of the relationships between two variable components. When the investigation is finished, a complete mathematical statement of the ecological process can be prepared in computer language. It then becomes possible to vary the values of the several parameters in the computer description and determine the results of such variation. This method has a high degree of precision. Reality is assured to the degree that the experimental animals respond as they do in nature. The degree of predictability is similarly contingent on the reality of the experimental conditions. In his investigations, Holling has switched from one experimental species to another. He has used mice, mantids, his technicians, and fish. This variety in experimental subjects reflects Holling's contention that in this fashion one deals purely with basic components and avoids species-specific individualities. However, at the same time one cannot identify the feed-back mechanisms contributing to stability in naturally evolved predator-prey complexes.

Slobodkin and Watt have both used inductive methods for the development of ecological generalizations. Slobodkin (1960) used his own experimental data on exploitation of a population of a prey species by a predator to derive generalized statements about the responses of the prey to various treatments. Watt (1968) advocated detailed statistical analyses of observational data to derive mathematical relationships between variables in a natural system so that these relationships could be incorporated into an elaborate computer simulation of the system. Thereafter, manipulation of the computer model would provide deductive information about the performance of the system under different conditions.

In Slobodkin's work both the precision and reality were high. Slobodkin discovered, however, that the relationships that appeared in two species did not exist in a third. The generality of the statements was, consequently, suspect. The precision and reality in Watt's mathematical computer models are, as Watt emphasized, determined by the accuracy of the methods used to collect the data and the skill in fitting mathematical expressions to them. No claim to generality is made for such models. Presumably generalizations might result from the repetitive appearance of similar statements in a succession of models.

In this brief review of some of the better known strategies for the elaboration of ecological theory we have seen that none is equally powerful for simultaneously producing the qualities of realism, generality, and precision. Some create general statements, some precise statements, some realistic statements, and most produce combinations of two of these. Against this background I should like to outline a method which seems to combine the most valuable components of various designs for investigation. It represents an approach that is being increasingly advocated (King and Paulik, 1967). The study by Hughes and Gilbert (1968) of the cabbage aphid is a good example.

One begins with the basic premise that the root source of generalizations is observations of organisms in nature. These observations should be performed by the investigator himself. If one is endeavoring to formulate broad statements which apply to a wide spectrum of species and communities, one also draws on the published observations of other biologists. If a sufficient variety of writings are sampled and if the information is adequately catalogued, it becomes possible to discern certain relationships that seem to appear again and again. Here the importance of Holling's idea of component analysis is realized. With the system analyzed into sub-processes, information from the literature can be assigned to one or more such categories. In this way the data are readily collated, and seeming similarities in response can be identified. Such similarities provide corroborative support for the initial hypothesis.

Regardless of the amount of information gathered or the degree of insight of the investigator, such hypotheses are invariably blurred, tentative, and indeterminate. The issues are clouded by an overlay of variation resulting from the biology of the individual species. The question becomes one of resolving the general from the particular. It is at this point that one of the basic assumptions of the approach appears. It is my conviction that if a response or a mode of reaction to environmental variation is in truth general, meaning that it applies to nearly all organisms, then it should be observable in protozoa and other simple animals. In practical terms, I would not argue that a phenomenon which could not be observed in protozoa was not general. What I do believe, however, is that if a phenomenon is observable in many other animals and is also present in protozoa, it is likely to be exhibited in a much clearer form than in more complex organisms. One does not have to contend with such complexities as sex ratios, life history stages, or learning in protozoa.

A second conviction which I hold is that verification of a hypothesis is not possible from observation alone. True, one may amass such an overwhelming degree of observational confirmation that for all practical purposes the hypothesis is supported. But for absolute conviction, one requires two things, experimental demonstration and accurate prediction. Further, quantification of a relationship is at best difficult from observation alone, and in many cases is impossible. Too many reactions are usually occurring simultaneously for the field investigation to provide clear-cut quantitative values. If one shares my enthusiasm for the importance of experimentation, then micro-organisms are ideal subjects. They are easily cultured in a small space and results of experiments are attainable in a few weeks.

To this point, then, I have suggested that the derivation of ecological generalizations should begin with observations in nature and proceed to the testing of resulting hypotheses by experimentation on simple organisms. These experiments provide two results: verification or rejection of the hypothesis, and quantitative measures of the relationships between the variables investigated.

Assuming the hypothesis is confirmed, the quantitative results provide the means for generating the second element needed for verification: successful prediction. The method used will, of course, vary from investigator to investigator and with the phenomenon being investigated. Computer models have two advantages. Their pseudo-algebraic languages are particularly adapted to descriptions of biological responses (Watt, 1966). Experimental manipulation of the model can predict future events if the model is a correct representation of a natural system (Paulik and Green-ough, 1966; King and Paulik, 1967; Conway, in press). The final stage, then, is a check of the accuracy of the prediction against events in the field. Needless to say, first efforts are not likely to be entirely accurate. The hypothesis is then either corrected, modified, or amplified and again verified by experimentation. The changes are then incorporated into the model and new predictions made.

With this brief description of the method, let us examine how well it meets the criteria of realism, precision, and generality. Its realism derives principally from the fact that original hypotheses are formulated from observations of phenomena in natural communities. Consequently, the investigator can be confident that he is indeed investigating a real rela-tionship rather than an imaginary or hypothetical one. Further, the re-quirement that predictions resulting from laboratory experiments and models shall be verified in the field provides additional assurance of the reality of the relationships. Occasionally the sequence of events may be reversed. Within the simplified conditions which exist in laboratory experiments or cultures, some interaction which might easily be obscured under natural conditions may be discovered. Armed with the insight provided by his laboratory observations, the biologist may then be able to discern the same relationship in the field. An example of such a process will be discussed further on.

The degree of precision that this method provides is largely a product of the precision of the field observation and the sophistication of the laboratory procedures used in the experiments. In the laboratory, given enough expensive equipment and enough time, one can achieve almost any degree of precision. The hazard here is not lack of accuracy but an excess of it. There exists an almost continual temptation to strive for an elegance of results that is relatively meaningless when these results are considered as models of events in the variable outside world. To an ecol-ogist, the required result is one which either confirms or denies a hypothe-sis. Effort expended beyond this point is diverted from the main goal.

For clarity, a comment must be made about these two words "precision" and "predictability." "Precision" is used here in the sense of accuracy of observations or measurements in the field or laboratory. Additionally, it may mean the accuracy of statements or equations in a model as meas-ured by the closeness of predicted to observed results in laboratory ex-periments. "Predictability" is measured by the capacity to provide quan-titative estimates of the values of parameters of field populations at future dates. It is here that the value of both the laboratory experiments and the model constructed from their results is tested. If the predictions fail, then the predictability is poor regardless of the precision of the sub-processes by which the predictions were produced.

Generality is at the same time the most valuable and the most difficult quality to achieve. As others have pointed out, there is a fundamental antithesis between precision and reality on the one hand, and generality on the other. Broad, non-specific generalities are easy to formulate. "No two species having the same requirements can co-exist in the same niche indefinitely." One need only set a mammal trap line through a meadow or look at a plankton sample to become immediately aware of the divergence of that simplistic statement and reality. On the other hand, if one is to be precise, he must be precise about something, in this case a particular organism or group of similar organisms. How is one to know whether the relationships established are common to all organisms and not just to those under study?

There are two obvious means of establishing generality. One is to argue, as I have, that any relationship exhibited by primitive organisms is probably found in all organisms. This is the keystone in the outlook of cell and molecular biologists. It is, however, an unverified assumption, an article of faith, and not a fact. The other means is to wait until the relationship has been identified in all other organisms. This is impossible as well as being excessively negative. The happy medium is to establish one's generalizations on one group of animals and then verify their applicability to one or more dissimilar groups of organisms. If the generalizations hold for the others, it is reasonable to presume that they will apply to most, if not all, other organisms. In my own case, I am involved either personally or via my graduate students in the study of predation in protozoa, rotifers, cladocerans, fish, and birds. If the ideas we generate are found to apply equally well to all these groups, it seems safe to assume that they have a reasonable degree of generality.

What are the shortcomings of this procedure which I am espousing? There are two. First, it is slow and time consuming. One can perform an enormous amount of investigation before producing one prediction, and that one may be wrong. Second, it demands a wider than normal span of skills in the investigator. He must be capable of field study of his animals, create discriminating laboratory experiments, and formulate computer models, which demands programming ability. These requirements may seem unreasonable, but in fact they are not. Most laboratory experimentalists are avid field biologists of their experimental animals. Computer manipulation is somewhat like a foreign language. If one learns it young enough it is not difficult. For someone of what is usually described as mature years it can be a real effort, as I can attest. But I have faith that even older persons can become fluent to the degree necessary for the needs outlined here.

Inasmuch as I am advocating the foregoing strategy, I naturally seek to follow it in my own researches. I should like to report, then, the present state of my investigation of a simple system composed of a straight food chain consisting of bacterial substrate, *Aerobacter*, *Paramecium*, and *Didinium*.

Field populations of *Paramecium* and *Didinium* are by no means as common as elementary textbooks would lead one to believe. However, Allen (1920) found them consistently in the Stockton Channel during the winter months. Densities of *Didinium* varied from essentially zero to highs of about 150,000 animals per cubic meter. My own field collections have produced samples with *Didinium* densities somewhat higher than this, the maximum being 180,000 per cubic meter in samples from a permanent slough running through sheep pastures. *Paramecium* densities in the Stockton Channel for all species combined likewise varied from very low to highs of 580,000 per cubic meter. Hairston (1967) in Michigan found *Paramecium aurelia* populations to have two peaks per year. We do not as yet have a series of observations through several seasons. Presumably, however, populations of both *Didinium* and *Paramecium* appear and disappear in the fauna of these bodies of water numerous times during the course of one season as well as from one year to the next. By disappear I mean to fall to such low densities as to be recorded as zero in any sampling program.

When more detailed information from the field is available, we will be able to comment on two questions basic to any laboratory investigation. First, what are normal densities in natural populations? One of the frequent criticisms leveled at laboratory population ecologists is that the animals are performing at unnaturally high densities, ones which would never be experienced in the field. The other answer concerns the frequency and amplitude of the oscillations in number of bacteria, *Paramecium* and *Didinium*. It cannot be stated with certainty, but Allen's data suggest that natural systems of these three kinds of organisms do not have a single pulse for each element after which it goes to extinction as is the common result in laboratory populations. If this does not occur in nature, the question is, why not? I shall return to this question further on.

Laboratory investigations are designed to provide measurements of interrelationships to be used in a computer simulation model. For brevity, then, I shall list here the components which appear at this time to be required for a simulation model of the *Aerobacter-Paramecium-Didinium* system. Values for most of the components are available as either qualitative or quantitative measurements. Many of the measurements are from the literature, especially in the reports by Beers. We are currently in the process of measuring the additional values needed.

AEROBACTER–PARAMECIUM–DIDINIUM SYSTEM

Aerobacter Density Component

1. Fission Rate
   a. Substrate supply rate
   b. *Aerobacter*
2. *Paramecium* Feeding Rate
   a. *Paramecium* density
   b. *Aerobacter* density
   c. *Paramecium* swimming rate

Paramecium Density Component

1. Fission Rate
   a. *Paramecium* density
   b. Feeding rate
      1. *Aerobacter* density
      2. Swimming rate
      3. *Paramecium* density
      4. *Didinium* density
2. *Didinium* Feeding Rate

Didinium Density Component

1. Fission Rate
   a. *Didinium* density
   b. Feeding rate
      1. Collision frequency
         a. Swimming rate
            1. *Didinium* density
            2. Hunger
            3. *Paramecium* density
            4. *Aerobacter* density
         b. *Didinium* size
         c. *Paramecium* size
         d. *Paramecium* swimming rate

    2. Attack as percentage of collisions

        a. Hunger

    3. Success in attack as percent of attacks

  c. Food quality—nutritional state of *Paramecium*

2. Mortality Rate

  a. Encystment

    1. *Paramecium* density
    2. Nutritional state of *Paramecium*
    3. *Didinium* density
    4. Bacterial activity
    5. Rate of fall of *Paramecium* density

        a. *Didinium* feeding rate
        b. *Didinium* density
        c. Bacterial density

    6. Time since conjugation in generations

  b. Death rate

    1. *Paramecium* density
    2. Starvation time

3. Recruitment Rate

  a. Excystment

    1. Time since encystment
    2. Bacterial density
    3. Cyst density

  b. Colonization

Interference Component

  1. Species Diversity
  2. Pyramid of Numbers
  3. Geometric Diversity

Some special comment is required about the interference component listed. It is quite apparent that in nature no predator could afford to exterminate its prey nor any herbivore eliminate its food resource. Yet in laboratory model populations predators commonly do just that. *Di-*

dinium in laboratory cultures invariably exterminate the *Paramecium* population even when bacteria are supplied for *Paramecium* food. Gause (1934) in his classical experiments was only able to maintain interacting populations by providing a refuge in the bottom of the dish for the *Paramecium* and by periodically inoculating with *Didinium* when the prior individuals had died out.

Given the right conditions, however, *Didinium-Paramecium* populations may persist in active form in laboratory cultures for extended periods of time. In my laboratory I have a small 60 mm petri dish containing about 8 ml. of fluid in which *Didinium* and *Paramecium* have persisted in fluctuating numbers for over a year. This culture differs from others in that it contains a wheat grain on which is growing an exuberant flora of phycomycetes. They form a dense fuzz around the wheat germ. *Paramecium* penetrate deeply into this layer, but *Didinium* rarely go in beyond the surface layer. Thus a certain percentage of the *Paramecium* is exempt from attack due to the interference of the fungi with the hunting of the *Didinium*.

I have attempted to duplicate these conditions with abiotic materials, to date without success. I have tried plastic straws, millimeter glass tubing, fine capillary tubing, glass wool, and miniature cotton balls tied with thread. None of these substances provide crevices small enough. However, these efforts have suggested two ideas. The first is that the refuge must not only provide spaces of the right dimensions for the prey relative to the predator, but it must also have some attractant which will draw a portion of the prey population into it. In the case of the wheat grain, the bacterial concentration is presumably higher in the immediate vicinity of the grain.

The second idea is a product of my desire to produce a "clean" experimental system. I wanted a non-biological substance for the complexity so that I could provide either bacteria or substrate for the bacteria in known amounts. This led to a mental distinction in natural communities between biotic and abiotic or physical complexity. Further reflection has convinced me that practically all the complexity in natural communities which one tends to think of as physical is, in reality, biotic. To a Cooper's hawk chasing a bird through a forest, a tree trunk is a physical obstacle which it must fly around. Another large animal such as a bear or a deer would similarly be a physical obstacle. Yet all these are living organisms. In some communities there may be true physical complexity; silt in a river, for instance, or rocks and crevices in the intertidal zone. But in the majority of cases the complexity of the habitat is provided by other species which do not directly enter into the living processes of the organism. It is for this reason that I believe that biotic complexity and habitat complexity may be regarded as the same thing. At this stage in the investigation it is not possible to state how one quantifies this component nor how to measure its impact on the processes within a given ecological subsystem such as the one I am discussing here. Foliage height diversity is one method useful in the study of avian faunas (MacArthur and MacArthur, 1961; MacArthur, MacArthur, and Preer, 1962).

In the specific system that I am investigating, any organisms present other than *Paramecium* represent obstacles to a *Didinium*. In its random swimming it will collide with them, but will derive no food benefit from the encounter. The rate of such collisions is the product of the number and size of these other organisms. Similar remarks apply to bacteria that *Paramecium* will not eat. They will be caught in the esophagus and must be rejected, which takes time. During that time, the *Paramecium* cannot be capturing an edible bacterium. Fine silt particles in water is another example. This response is particularly easy to observe in free-swimming rotifers in which the corona is withdrawn when an inedible object is touched. These kinds of interference in free-swimming micro-organisms can be quantified. In the study of complex organisms more elaborate measures of the interference component must be developed.

Summary

The title of this paper could easily have been phrased as a question: What is the role of laboratory experimentation in ecological research? To summarize the foregoing views, it is my conviction that laboratory experiments form one of the three basic procedures that an individual must use in the investigation of ecological phenomena.

Literature Cited

Allen, W. E. 1920. A quantitative and statistical study of the plankton of the San Joaquin River and its tributaries in and near Stockton, California, in 1913. Univ. Calif. Publ. Zool. 22: 1-292.

Conway, G. R. In press. Computer simulation as an aid to developing strategies for anopheline control. Misc. Publ. Entomol. Soc. Amer.

Evans, F. R. 1958. Competition for food between two carnivorous ciliates. Trans. Amer. Microsc. Soc. 77: 390-395.

Garfinkel, D. 1967. A simulation study of the effect on simple ecological systems of making rate of increase of population density-dependent. J. Theoret. Biol. 14: 46-58.

Gause, G. F. 1934. The struggle for existence. Williams and Wilkins Co., Baltimore.

Hairston, N. G. 1967. Studies on the limitation of a natural population of *Paramecium aurelia*. Ecology 48: 904-910.

Hairston, N. G. and S. L. Kellermann. 1965. Competition between varieties 2 and 3 of *Paramecium aurelia:* The influence of temperature in a food-limited system. Ecology 46: 134-139.

Hairston, N. G., J. D. Allan, R. K. Colwell, D. J. Futuyma, J. Howell, M. D. Lubin, J. Mathias, and J. H. Vandermeer. 1968. The relationship between species diversity and stability: an experimental approach with protozoa and bacteria. Ecology 49: 1091-1101.

Holling, C. S. 1963. An experimental component analysis of population processes. Mem. Entomol. Soc. Can. No. 32: 22-32.

——————. 1966a. The functional response of predators to prey density. Mem. Entomol. Soc. Can. No. 48: 1-86.

——————. 1966b. The strategy of building models of complex ecological systems. pp. 195-214. *In* K. E. F. Watt (ed.) Systems analysis in ecology. Academic Press, New York and London.

Huffaker, C. B. 1958. Experimental studies on predation. II. Dispersion factor and predator-prey oscillations. Hilgardia 27: 343-383.

——————, K. P. Shea and S. G. Herman. 1963. Experimental studies on predation. Complex dispersion and levels of food in an acarine predator-prey interaction. Hilgardia 34: 305-330.

Hughes, R. D. and N. Gilbert. 1968. A model of an aphid population—a general statement. J. Anim. Ecol. 37: 553- 563.

Hutchinson, G. E. 1957. Concluding remarks. Cold Spring Harbor Symp. Quant. Biol. 22: 415-427.

——————. 1967. A treatise on limnology. Vol. II. Introduction to lake biology and the limnoplankton. John Wiley and Sons, Inc. New York.

King, C. E. and G. J. Paulik. 1967. Dynamic models and the simulation of ecological systems. J. Theoret. Biol. 16: 251-267.

Levins, R. 1968. Evolution in changing environments. Princeton Univ. Press, Princeton.

MacArthur, R. H. 1955. Fluctuations of animal populations and a measure of community stability. Ecology 36: 533-536.

——————. 1965. Patterns of species diversity. Biol. Rev. 40: 510-533.

—————— and J. W. MacArthur. 1961. On bird species diversity. Ecology 42: 594-598.

——————, J. W. MacArthur and J. Preer. 1962. On bird species diversity. II. Prediction of bird census from habitat measurements. Amer. Nat. 96: 167-174.

Margalef, R. 1968. Perspectives in ecological theory. Univ. Chicago Press, Chicago.

Miller, R. S. 1967. Pattern and process in competition. pp. 1-74. *In* J. B. Cragg (ed.) Advances in Ecological Research. Vol. 4. Academic Press, New York.

Morse, D. H. 1967. Foraging relationships of brown-headed nuthatches and pine warblers. Ecology 48: 94-103.

Munro, J. 1967. The exploitation and conservation of resources by populations of insects. J. Anim. Ecol. 36: 531-547.

Odum, E. P. 1969. The strategy of ecosystem development. Science 164: 262-270.

Paulik, G. J. and J. W. Greenough, Jr. 1966. Management analysis for a salmon resource system. pp. 215-252. *In* K. E. F. Watt (ed.) Systems analysis in ecology. Academic Press, New York and London.

Preston, F. W. 1962. The canonical distribution of commonness and rarity. Ecology Part I, 43: 185-215; Part II, 43: 410-432.

Slobodkin, L. B. 1960. Ecological energy relationships at the population level. Amer. Nat. 94: 213-236.

Sluss, R. R. 1967. Population dynamics of the walnut aphid, *Chromaphis juglandica* (Kalt), in northern California. Ecology 48: 41-58.

Watt, K. E. F. 1966. Systems analysis in ecology. Academic Press, New York and London.

—————————. 1968. Ecology and resource management. McGraw-Hill Book Co., New York.

Frieda B. Taub

A Continuous Gnotobiotic
(Species Defined) Ecosystem

Abstract

The responses of autotrophic communities to change in light intensity, nutrient limitation, and dilution rate were studied by means of a two-stage continuous-culture apparatus. The communities consisted of the single alga *Chlamydomonas reinhardtii*, the protozoan *Tetrahymena vorax*, and the bacteria *Pseudomonas fluorescens* and *Escherichia coli*; 0.5 mM NO_3^- was the limiting nutrient. At high dilution rates (1.0 and 1.4 per day) algal populations and *Pseudomonas* were low in the upstream community, and the protozoa and *E. coli* became extinct, whereas at lower dilution rates (0.57 and 0.73) all of the organisms were able to maintain steady-state populations. Under reduced light intensities, and at the lower dilution rates, the algal populations attained a sparse steady state and the protozoan population was able to maintain a growth rate of 0.5 but not 0.75. Kinetic relationships between algal cell concentration and protozoan growth rates accurately predicted extinction, but not steady state populations of the protozoan.

The downstream and yield communities were progressively denser and more stable under environmental changes.

Introduction

The experiment reported here was designed to show the responses of an autotrophic community to changes in light intensity, nutrient limitation, and dilution rate. It represents an advance over our previously reported studies (Taub, 1969a; Taub, 1969b) in that continuous-culture methods provided for continuous growth during the duration of the study and allowed the estimation, not only of standing crop, but also of net growth rates and productivity rates during steady state conditions. The use of a "gnotobiotic" (Dougherty, 1959) community permitted enumeration of the populations of all component species and minimized uncontrolled variables.

The model falls within Patten's (1968) definition of a system " . . . an assemblage of objects united by some form of interaction or interdependence in such a manner as to form an entirety or whole." Interaction between the organisms was assured by the fact that the heterotrophs depended on the autotrophs for their source of organic compounds, and the protozoa could ingest both the algal cells and the bacteria in addition to using extracellular products. The system was used as a model of a real biological community in the sense of Patten (1968); as a model, it is an abstraction and a simplification, but not a substitute for a real system. The behavior of the model must not be inferred to apply to real communities unless support for such an inference is supplied. In this study the model is used almost entirely as an analytical tool.

Previous work on microbial ecosystems has been reviewed by Dennis Cooke in the symposium.

Materials and Methods

The organisms consisted of the alga *Chlamydomonas reinhardtii*, Strain 90, of the Indiana culture collection, the protozoan *Tetrahymena vorax*, Strain V_2S, from G. G. Holtz, Jr., and the bacteria *Pseudomonas fluorescens*, 13525, and *Escherichia coli*, 14948, both of the American Type Culture Collection. The chemical environment was Medium 63 (Taub & Dollar, 1968a). This medium is composed of reagent grade inorganic chemicals, with the exception of 0.0284 mM EDTA, and does not support significant growth of the heterotrophic organisms but supplies all of the requirements for autotrophic growth of *Chlamydomonas*. The continuous-culture apparatus consisted of a) a medium reservoir, which delivered aseptic medium by means of a digital pump; b) a first (upstream) growth flask, the overflow of which was delivered to c) a second (downstream) growth flask, the overflow from which was delivered to d) a yield reservoir. The flow into the first flask was determined by a digital metering pump (the medium remaining within closed Tygon tubing). Since the volume of liquid in each flask was maintained constant by an overflow mechanism, the volume delivered to the yield bottle represented the flow into each of the growth flasks. Mixing was achieved both by aeration with a 2% CO_2-enriched air mixture and rocking at 17°, at 20 cycles per min. Loose glass beads on the bottom of each flask were used to dislodge cells that adhered to the sides of the flasks. The flask is described in detail in Taub & Dollar, 1968 and 1965. Temperature was controlled at 20C± 1 degree, and lights were supplied by combinations of fluorescent and incandescent bulbs in a Sheer growth unit. Light intensity was measured by a GE 213 light meter from which the filter had been removed. Therefore these values were approximately twice as high as would have been attained had the filter been used. (Removal of the filter in part eliminated the relative insensitivity of the meter to wave lengths less than 5,000 Å or greater than 7,000 Å). Two replicate systems designated green (system 1) and red (system 2) were studied. The only recognized difference between the systems was a greater volume in flask 1 of the green system, caused by a crack in the overflow tube; volumes are shown in Figs. 1 and 2.

Chlamydomonas were enumerated by hemacytometer, *Tetrahymena* by the dilution method of Sonneborn (1950), and the bacteria by plating on King's Medium B (King, et al., 1954), and EMB media (Difco, 1953). The lowest detectable populations for these methods were 1×10^4 *Chlamydomonas*/ml, 5 *Tetrahymena*/ml, and 10^2 bacteria/ml.

Pigment measurements were conducted in accordance with the methods of Strickland and Parsons (1965) and the use of a 1-cm cuvette and the P.S. formula. Analyses of NO^-_3 and NO^-_2 were determined by the method of Wood et al. (1967) and Bendschneider and Robinson (1952). Oxidizable carbon was determined by a modification of the method of Strickland and Parsons (1965); whole samples rather than particulate matter were analyzed since Medium 63 did not interfere as would sea water.

All of the organisms were enumerated into the first flask on day 0 and permitted to grow as batch cultures until day 5. A flow of approximately 750 ml per day was initiated on day 5 and maintained until day 13, at which time the flow was decreased to approximately 400 ml/day. Light intensity was maintained at approximately 2,000 ft-c until day 43, when it was reduced to half that intensity. Optical density was measured daily in each flask by use of the nephelometer tube. Samples were aseptically drawn from each of the flasks and yield reservoir for species enumeration two or three times weekly. The chemical analyses were largely limited to samples from the yield flask. Yield was removed daily.

Results and Discussion
Standing Crop Relationships

The light absorbance and densities of the organisms are shown in Figs. 1 (green replicate) and 2 (red replicate). During the first five days of batch growth, only the algal population increased rapidly. Densities and exponential growth rate ($\mu = 2.3 \dfrac{(\log x_t - \log x_0)}{t}$) are shown in Table 1. When high flow was initiated (Table 2), the algal and *Pseudomonas* populations remained stable, whereas the protozoa and *E. coli* washed out. When the flow was reduced (Table 3), algal concentrations increased and assumed a steady state, the *Pseudomonas* populations stayed at the same density for almost two weeks and then increased in both replicates, and the protozoan and *E. coli* populations rose from undetectable levels to low, steady states. When the light intensity was reduced on day 43

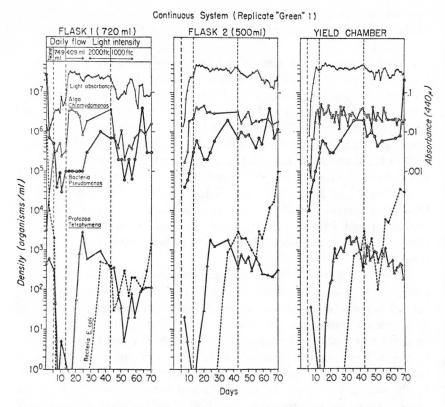

FIGURE 1. Population densities and light absorbances for the three communities of the green system. (System 1).

(Table 4), the algal population decreased to a new steady state; the protozoan population in the green replicate decreased to 2×10^2 and that in the red replicate became undetectable; the *Pseudomonas* population decreased slightly; and the *E. coli* population increased slightly. A comparison of the growth rates during initial batch growth with those during the dilution phases indicates that algal growth was not at its maximum and there was no growth on the part of the other organisms, probably as a result of residual lag stages.

The downstream communities received their population upon initiation of overflow from the upstream community on day 5. In general, all populations were slightly denser and showed less fluctuation than the upstream community did; e.g., in the upstream community, the algal populations changed from ca. 2.9×10^5 to 2.1×10^6 to 7.6×10^5 cells/ml during the three time periods (a sevenfold maximum change), whereas in the downstream communities the changes were only 1.1 to 3.2 to 2.1

TABLE 1.—Growth Characteristics During Batch Growth.

	DENSITY		Exponential (ln) Growth Rate (per day) μ
	Initial	Day 5	
System 1 (green)			
O.D.	.015	.03	.14
Alga	3.3×10^3	2.7×10^5	.88
Protozoan	4.5×10^2	3.1×10^2	$-.08$
Pseudomonas	2.8×10^7	4.5×10^5	$-.825$
E. coli	1.0×10^6	2×10^3	-1.29
System 2 (red)			
O.D.	.012	.028	.17
Alga	3.3×10^3	1.7×10^5	.78
Protozoan	4.5×10^2	1.5×10^2	$-.22$
Pseudomonas	2.8×10^7	7×10^5	$-.74$
E. coli	1×10^6	1×10^3	-1.38

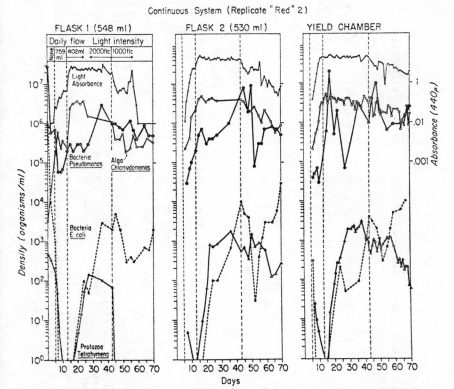

FIGURE 2. Population densities and light absorbances for the three communities of the red system. (System 2).

TABLE 2.—Density, Growth Rates and Production Rates of Organisms, High Light, High, High Dilution Rate, Days 7 to 13 Inclusive.

SYSTEM 1

Avg. flow 749 ml/day, Avg. dilution rate for Flask 1 = 1.040, Flask 2 = 1.497, Avg. light intensity 2,000 ft.-c.

	Standing Crop (cells per ml)			Exponential (ln) Growth Rate (per day)			Production (cells per day)		
	Flask 1	Flask 2	Yield	Flask 1	Flask 2	Yield	Flask 1	Flask 2	Yield
O.D.	.05	.16	.27	1.04	1.06	.48	3.5×10^1	1.2×10^2	2.0×10^2
Alga	2.9×10^5	1.1×10^6	1.74×10^6	1.04	1.12	0.44	2.1×10^8	8.4×10^8	1.3×10^9
Protozoan	(2.5)	6.3	7.0	(1.04)	0.90	.11	(1.9×10^3)	4.7×10^3	5.2×10^3
Pseudomonas	6.3×10^4	1.1×10^5	1.2×10^5	1.04	0.65	.10	4.7×10^7	8.2×10^7	9.1×10^7
E. coli	0	(1.0×10^{-5})	(1.0×10^{-5})	0	0	0	0	0	0

SYSTEM 2

Avg. flow 759 ml/day, Avg. dilution rate for Flask 1 = 1.379, Flask 2 = 1.431, Avg. light intensity 2,000 ft.-c.

	Standing Crop (cells per ml)			Exponential (ln) Growth Rate (per day)			Production (cells per day)		
	Flask 1	Flask 2	Yield	Flask 1	Flask 2	Yield	Flask 1	Flask 2	Yield
O.D.	.05	.01	.02	1.38	0.93	.60	4.3×10^1	1.2×10^2	2.2×10^2
Alga	3.7×10^5	1.1×10^6	1.7×10^6	1.38	0.95	.41	2.8×10^8	8.3×10^8	1.3×10^9
Protozoan	(1.25)	(1.25)	8.0	(1.38)	(0)	1.85	(9.5×10^2)	(9.5×10^2)	6.1×10^3
Pseudomonas	9.6×10^4	1.1×10^5	8.4×10^4	1.38	0.19	−.28	7.3×10^7	8.4×10^7	6.4×10^7
E. coli	0	(1.0×10^{-5})	(1.0×10^{-5})	0	1.43	0	0	(7.6×10^{-3})	(7.6×10^{-3})

( ) = extinct for practical purposes, average less than detectable limits.

TABLE 3.—Density, Growth Rates and Production Rates of Organisms, High Light, Low Dilution Rate, Days 24 to 43 Inclusive.

SYSTEM 1

Avg. flow 409. ml/day, Avg. dilution rate for Flask 1 = 0.568, Flask 2 = 0.818, Avg. light intensity 2,000 ft.-c.

	Standing Crop (cells per ml)			Exponential (ln) Growth Rate (per day)			Production (cells per day)		
	Flask 1	Flask 2	Yield	Flask 1	Flask 2	Yield	Flask 1	Flask 2	Yield
O.D.	.24	.39	.42	0.57	0.32	.08	9.8×10^1	1.6×10^2	1.7×10^2
Alga	2.1×10^6	3.2×10^6	3.0×10^6	0.57	0.29	−.06	8.5×10^8	1.3×10^9	1.2×10^9
Protozoan	1.2×10^3	1.3×10^3	1.4×10^3	0.57	0.07	.06	4.9×10^5	5.4×10^5	5.7×10^5
Pseudomonas	5.3×10^5	8.8×10^5	1.4×10^6	0.57	0.33	.43	2.2×10^8	3.6×10^8	5.5×10^8
E. coli	2.3×10^2	9.8×10^2	9.4×10^2	0.57	0.63	−.04	9.2×10^4	4.0×10^5	3.84×10^5

SYSTEM 2

Avg. flow 402. ml/day, Avg. dilution rate for Flask 1 = 0.730, Flask 2 = 0.758, Avg. light intensity 2,000 ft.-c.

	Standing Crop (cells per ml)			Exponential (ln) Growth Rate (per day)			Production (cells per day)		
	Flask 1	Flask 2	Yield	Flask 1	Flask 2	Yield	Flask 1	Flask 2	Yield
O.D.	.24	.43	.46	0.73	0.33	.06	9.7×10^1	1.7×10^2	1.8×10^2
Alga	2.1×10^6	3.8×10^6	3.5×10^6	0.73	0.34	−.08	8.6×10^9	1.5×10^8	1.4×10^9
Protozoan	8.4×10^1	9.9×10^2	1.7×10^3	0.73	0.70	.52	3.4×10^4	4.0×10^5	6.7×10^5
Pseudomonas	1.0×10^6	2.0×10^6	1.4×10^6	0.73	0.37	−.36	4.0×10^8	7.9×10^8	5.5×10^8
E. coli	8.4×10^2	2.7×10^3	1.2×10^3	0.73	0.52	−.82	3.4×10^5	1.1×10^6	4.8×10^5

TABLE 4.—Density, Growth Rates and Production Rates of Organisms, Low Light, Low Dilution Rate, Days 45 to 70 Inclusive..

SYSTEM 1

Avg. flow 402. ml/day, Avg. dilution rate for Flask 1 = 0.558, Flask 2 = 0.804, Avg. light intensity 1,220 ft.-c.

	Standing Crop (cells per ml)			Exponential (ln) Growth Rate (per day)			Production (cells per day)		
	Flask 1	Flask 2	Yield	Flask 1	Flask 2	Yield	Flask 1	Flask 2	Yield
O.D.	.12	.26	.33	0.56	0.45	.22	4.6×10^1	1.1×10^2	1.3×10^2
Alga	7.6×10^5	2.1×10^6	2.4×10^6	0.56	0.51	.2	3.1×10^8	8.2×10^8	9.6×10^8
Protozoan	1.9×10^2	4.4×10^2	6.3×10^2	0.56	0.46	.36	7.6×10^4	1.8×10^5	2.6×10^5
Pseudomonas	6.2×10^5	1.1×10^6	3.6×10^6	0.56	0.35	1.18	2.5×10^8	4.4×10^8	1.4×10^9
E. coli	1.1×10^4	1.6×10^4	1.2×10^4	0.56	0.22	-.29	4.6×10^6	6.3×10^6	4.7×10^6

SYSTEM 2

Avg. flow 411. ml/day, Avg. dilution rate for Flask 1 = 0.747, Flask 2 = 0.775, Avg. light intensity 1,220 ft.-c.

	Standing Crop (cells per ml)			Exponential (ln) Growth Rate (per day)			Production (cells per day)		
	Flask 1	Flask 2	Yield	Flask 1	Flask 2	Yield	Flask 1	Flask 2	Yield
O.D.	.04	.16	.25	0.75	0.57	.41	1.7×10^1	6.7×10^1	1.0×10^2
Alga	3.6×10^5	1.8×10^6	1.9×10^6	0.75	0.62	.08	1.5×10^8	7.3×10^8	7.9×10^8
Protozoan	0	5.6×10^2	5.8×10^2	0.75	0.78	.04	0	2.4×10^5	2.4×10^5
Pseudomonas	7.4×10^5	2.2×10^6	2.7×10^6	0.75	0.52	.20	3.0×10^8	9.2×10^8	1.1×10^9
E. coli	2.4×10^4	5.9×10^3	3.3×10^3	0.75	-2.34	-.57	9.7×10^6	2.4×10^6	1.4×10^6

x 10^6 cells/ml (a threefold maximum change). The protozoan populations were especially more stable in the downstream communities. Once they became established during the low dilution period, they reached a steady density of ca. 1 x 10^3 cells/ml. They were little affected by the drop in light intensity and there was only a slight but constant downward trend.

The populations in the yield reservoirs were usually slightly denser and more stable than in the downstream community.

Growth Rates and Production Estimates

Assumptions and Their Validity

Unlike the data analyses above, the estimation of growth rates and production rates required various assumptions, outlined below. These have been adapted from the excellent review of Herbert (1964). Our system conformed to a single-stream, multi-stage continuous culture. Culture medium containing the growth-limiting nutrient, s, at a concentration, s_o, (NO^-_3, 0.5 mM) entered flask 1 at a flow-rate, f, and culture flowed from flask 1 to flask 2 and finally out of the latter at the same flow-rate, f. The dilution rates were respectively $D_1 = f/v_1$ and $D_2 = f/v_2$, where v is the culture volume. The concentrations of cells and of substrate in the two flasks were, respectively, x_1, x_2 and s_1, s_2. Growth was assumed to follow the exponential growth equation $\frac{dx}{dt} = \mu x$, where $\mu =$ exponential growth rate.

In the first flask, there was no inflow of cells, and the equation for cell balance is increase = growth — outflow; $\frac{dx_1}{dt} = u_1 x_1 - D_1 x_1$. Under steady state conditions, $\frac{dx_1}{dt} = 0$ and $u_1 = D_1$, i.e., the growth rate in the steady state was equal to the dilution rate. This condition is obvious when higher or lower growth rates are considered; the populations would increase or decrease so that a steady state would not prevail.

In the second flask, the equation for cell balance is:

increase = inflow — outflow + growth; $\frac{dx_2}{dt} = D_2 x_1 - D_2 x_2 + \mu_2 x_2$

At a steady state, $\frac{dx_2}{dt} = 0$ and $\mu_2 = D_2 \frac{(x_2 - x_1)}{(x_2)}$

These relationships assume perfect mixing and representative overflow.

The yield reservoir represents a special problem, since the population started from zero volume each day and accumulated gradually as the reservoir filled until it was replaced and sampled. Dilution did not occur as there was no overflow. On a population basis the exponential growth rate was calculated as:

$$\mu = 2.3 \ \frac{(\log x \text{ yield} - \log x_2)}{t}, \text{ where } t = 1 \text{ day.}$$

It may be argued that, on a physiological basis, individual cells were growing at twice this rate, since cells that entered at minute one had the full day to grow, whereas those cells that entered immediately before the replacement had no time to grow.

The relationship between limiting substrate and growth rate, which has been shown to hold in many cases, is described in the Michaelis-Menten relationship:

$$\mu = {}^{\mu} \max \frac{(s)}{K_s + s}, \text{ where } K_s = \text{"saturation constant,"}$$

numerically equal to substrate concentration at which $\mu = \frac{1}{2} \ \mu$ max.

This relationship has been shown by Monod (1942) to apply to bacteria in fermenters and has been demonstrated since by Jannasch (1967) and Hobbie & Wright (1965) to apply to natural populations of bacteria. The relationship has been verified for the alga *Chlorella* by Pearson, et al. (1969).

Monod (1942) also showed that the yield, Y, where

$Y = \frac{\text{weight of bacteria formed}}{\text{weight of substrate used}}$ is a constant (at least to a first approximation).

The doubling rate may be calculated from the exponential (ln) growth rate μ, by the relationship, doubling rate $= \frac{.693.}{\mu}$

The continuous-culture systems that utilize limited nutrient input as an external control on growth rate have been thoroughly reviewed by Novick (1955). He shows that the stability of the system is due to interaction between the population density and the nutrient concentration: the rising population density lowers the concentration of the required factor until the population growth rate slows. The growth rate decreases until it becomes equal to the washout rate. He also shows that the concentration of the controlling factor in the growth flask, s_1 or s_2, depends only on the growth rate (= flow rate) selected and is independent of the input concentration, s_0. (Notation changed to conform with Herbert's system.)

Production rate for each flask is $P = fx$, the flow rate multiplied by the density. Thus the production in the second flask is the result for the first and second flask, and that in the yield reservoir represents the yield of the total system. The yield produced within the second or yield chambers can be calculated by subtraction of the value of the previous chamber.

Before these assumptions were applied, consideration was given to their validity under the conditions that obtained.

Since the densities measured represent population levels after whatever consumption may have occurred, the growth and production rates are minimal, or net, rates. Gross growth and production would include those algal and bacterial cells that were produced and consumed before measurement.

Perfect mixing was probably not achieved, but the error would be small and was ignored. Growth on the chamber walls and tubes was minimized until the end of the experiment but did become dense enough by day 70 so that the experiment was terminated. Adherence of organisms on the walls would result in higher estimates of growth rates than had actually occurred since a larger population than measured would be contributing offspring. The surfaces may have served as a refuge for the protozoa and *E. coli* while they were undetectable in the liquid samples, although this is not necessary since the enumeration methods would not have shown very low populations.

The usual assumption that Y is constant and independent of s is not strictly valid for *Chlamydomonas*. In a nitrate-rich medium, the cells formed can be 2.5 times as high in N content (% of ash-free dry weight) as cells are in a medium in which the nitrate has been depleted (Taub, unpublished data). Also, the size and dry weight of the *Chlamydomonas* can vary at least twofold, and the *Tetrahymena* can vary more than fourfold (Taub, unpublished data). Therefore, the usual tacit acceptance of cell number as an estimate of substrate removed and biomass produced can be used only very approximately. (At the time of this study we did not have the capacity to measure cell volumes.)

So that the simplified formulae shown above could be used, it was necessary to assume steady state population densities. Periods of obvious rapid change, such as immediately after environmental changes, were not considered in these calculations, i.e., flow was initiated on day 5, but data were not used until day 7, days 14-24 were eliminated from consideration since the protozoa and *E. coli* populations were rapidly changing, and days 43-44 were eliminated from the calculations. Otherwise, the day-to-day changes were assumed to be due to experimental error, and all measurements within each of the three time periods were averaged (Tables 2-4). This method ignores the midperiod change in *Pseudomonas* density during the second period and the gradual downward trend of the protozoan density during the third period.

Upstream Community

It will be immediately noted that in the first flask the exponential growth rates were equal to the dilution rate so long as the population maintained itself (since we defined these populations as being in a steady state condition). These values indicate which organisms, under the stated conditions, were capable of growing at least at the dilution rate. For example, the loss of the protozoan and *E. coli* populations indicates that they could not maintain growth rates of 1.04 or 1.38 per day (high flow, high light, Table 2). Their growth rate could be calculated by a comparison of the washout rate with the dilution rate (Jannasch, 1969). Both of these organisms were able to grow at a rate of .57 and 0.73 per day when dilution was lower and algal density higher (low flow, high light, Table 3). *E. coli* could also grow at a rate of 0.75 per day when light intensity was reduced and the algal concentration decreased (Table 4). In contrast, the protozoan population could grow at 0.56 but not 0.75 under the reduced light and decreased algal cell concentration. Since we have no evidence that the change in light intensity directly influenced the protozoan population, its limitation is presumably related to the algal density.

The limitation of protozoan density and growth rate poses the question of limiting factor(s). Since this protozoan reaches population densities of 10^5 cells/ml in rich organic media, no direct density-dependent limitation is postulated. The possible substrates for protozoan nutrition were (1) algal cells, (2) algal extracellular products, and (3) bacteria. Protozoa were cultured on algal cultures with densities ranging from 8.4 to 3.0 x 10^6 cells/ml, and a μ_{max} of 1.1 and a $K_s = 0.7$ x 10^6 cells/ml were obtained (algal cell concentrations 0.57 and 0.16 x 10^6 cells/ml did not conform to the Michaelis-Menten line). On the assumption that either the algal cells were the sole nutrient source, or that extracellular products were proportional to cell concentration, the kinetic relationships defined above might be expected to explain protozoan behavior. The dilution rate, the algal standing crop, and the calculated growth rates are compared in Table 5. Extinction did occur in all three cases, as predicted by the kinetics. A steady-state population did occur in the one case where it was predicted. However, in the two cases where increasing populations were predicted, the populations were actually steady state. Theory would predict that the protozoan populations would have increased to the point where they cropped the algal population down to the density where the protozoan growth rate was reduced to the dilution rate. This did not occur. So that it could be determined whether extracellular products were a significant source of nutrient for protozoan growth, an experiment was conducted in which the algal concentration

TABLE 5.—Comparison of Actual Protozoan Population Behavior with That Predicted by Kinetics.*

	Algal concentration x10^6 cells/ml	μ	D	Predicted	Actual	Cells/ml
System 1 (days)						
7–13	0.29	0.31	1.0	extinction	extinction	
24–43	2.1	0.82	0.57	increasing pop.	steady state	1.2x10^3
45–70	0.76	0.56	0.56	steady state	steady state	1.9x10^2
System 2 (days)						
7–13	0.37	0.37	1.38	extinction	extinction	
24–43	2.1	0.82	0.73	increasing pop.	steady state	8.4x10^1
45–70	0.36	0.36	0.75	extinction	extinction	

$$^*\mu = \mu_{max} \frac{s}{K_s + s} \qquad K = 0.73 \times 10^6 \text{ algal cells/ml}$$
$$\mu_{max} = 1.1$$

was varied from nil to ca. 4 x 10^6 cells/ml while extracellular material was kept constant. In the absence of algal cells, the protozoa showed no significant growth, and growth was directly related to cell concentration. These results suggest that the extracellular products either by themselves, or with a limited concentration of algal cells, do not form a significant source of nutrient. The importance of the bacteria as a nutrient source is difficult to evaluate. The protozoan does not require the presence of bacteria to grow on algae; the kinetic values were obtained from cultures in which no bacteria were known to be present. Also the presence of *E. coli* in the experiments described here does not correlate with the success of the protozoa since the *E. coli* populations increased under reduced lighting, whereas the protozoan populations diminished greatly in one case and became extinct in the other (Table 4). Nor did the *Pseudomonas* densities relate to protozoan populations. Of course the bacterial generation times could have varied and thus provided more cells during certain periods.

The absence of significant cropping of the algal population with the expected predator-prey cycles also deserves consideration. As in the previous experiments (Taub, 1969; Taub, 1969b), no significant cropping of algal populations could be demonstrated as long as light was supplied. When the protozoan is introduced into algal cultures and incubated in the dark for 4 days, 50-80% of the algal cells disappear, and Chlorophyll *a* and ^{14}C uptake ability are significantly decreased (Taub, unpublished). By variation of the concentration of NO^-_3, the μ_{max} of 2.0

and K_s of 0.16 mM were obtained for the *Chlamydomonas* under the high-light conditions (12-hour light-dark cycle). Since the μ_{max} of the protozoan when consuming algal culture was 1.1, it is suggested that the algal cells are able to regrow as fast as they are cropped provided light is available (these μ_{max} and K_m values should be accepted only as approximate and preliminary values). Predator-prey oscillations were shown for limited time periods in a continuous culture of the predator (a Myxameba, *Dictyostelium descoideum*) and prey, (*E. coli*) with a limiting substrate of glucose (Drake et al., 1966).

During the recovery of the protozoan populations in the first flask, the calculated growth rates were 1.56 per day and 1.17 per day by means of the equation $\mu = D + \frac{1}{t} \ln (x/x^0)$ from Jannasch (1969) and on the basis of days 20-24 for the green system and days 20-27 for the red system, i.e., the first day when a measurable population occurred, until population increase ceased). These growth rates were the fastest seen during the entire experiment, and are faster than the μ_{max} calculated from the other alga-protozoan experiment. (Higher μ_{max} are found when rich organic mixtures, such as yeast extract proteose-peptone solutions, are used.) The growth rates would have to have been greater than the dilution rates, since the population density rapidly increased in spite of losses from dilution. The substrate for this increased growth rate must be related to the increased algal population since no other major changes occurred (the increase in *E. coli* population cannot explain the increased protozoan growth since the bacterium did not appear in the green system until after the protozoan increase, and a high *E. coli* population did not prevent extinction of the protozoan population during low light intensity in the red system). If such high growth rates of the protozoa can be explained by an increase of algal concentration from ca. 3×10^5 to 2×10^6 cells /ml, why did the protozoan population cease to grow at greater than the dilution rate after day 27 when no significant decrease in algal cells occurred, and why did protozoan growth rates slow down in the second flask and yield reservoirs, where the algal densities were even higher?

Downstream Community

Under conditions of high dilution, algal growth rates were relatively high in the second flask. In the system 1 replicate, the unusual phenomenon of higher growth rates in the second flask than in the first can be explained by its higher dilution rate (since its volume was smaller). Presumably a relatively high amount of substrate carried over into the second flask because of the limited production in the upstream community (this is supported by the nitrate concentration remaining even in the yield. (Table 6.)) Growth and production rates for the other organisms were not consistent between replicates.

Under reduced flow, the algal growth and the new production in the second flask were relatively less. Presumably, the higher production in the first flask used up more nutrient and more sharply limited nutrient availability. The protozoan populations behaved differently in the two replicates; in the first, almost no further growth occurred; in the second replicate, growth continued at the same rate as in the first flask. In both, bacteria had significant growth rates.

Under reduced lights, the algal populations continued to grow almost as rapidly in the second flask as they had in the first. Presumably, the low production in the first flask also permitted the carryover of nutrient. In both replicates, protozoan growth rates and production rates were high. Growth rates for the *Pseudomonas* and the *E. coli* in the first replicate were significant, but showed a sharp decrease for *E. coli* in the second replicate.

Correlation of the growth rates between the organisms within a flask must be attempted with great caution since the dilution factor, D, is the same for all organisms in the flask, and D occurs within the calculation for growth rate $_{(\mu \, = \, D \, \dfrac{(x_2 - x_1)}{x_2}}$

Yield Community

Where algal growth rates were relatively high in the second flask (high dilution, Table 2, or low light, Table 4), slower, but significant growth continued in the yield reservoir. In the former case, nitrate was still available in the yield flask. Where growth was reduced in the second flask (high light, low dilution), slight losses occurred in the yield reservoir (Table 3). No appreciable nutrient was available (Table 6).

It is interesting to postulate whether the algal density would have shown significant losses if the condition had prevailed for a longer period of time, i.e., if there had been another growth flask in the series. The decline, if it was not within experimental error, might be due to lack of growth since there was no available nutrient and there was predation. However, since the protozoan population was not increasing its biomass, its consumption would have resulted in the release of their waste products equal to their consumption. Theoretically, the system could have reached a steady state provided that no significant quantity of products were refractory to recycling.

TABLE 6.—Average Yield Reservoir Concentrations in Two Systems Under Three Conditions of Dilution and Light Intensity.

	Dry wt, mg/ml	Ash wt, mg/ml	Biomass* (ash-free dry wt), mg/ml	Cox, mg/ml	Nkj, mg/ml	C/N	NO−3, mM	NO−2, mM	chl a mg/l	chl b mg/l	Car., mg/l	Car./chl a
High dilution, high light												
System 1	.357	n.d.†	—	.028	.006	4.7	0.87	0.000	.191	.099	.064	.34
System 2	.363	n.d.†	—	.031	.006	5.2	0.75	0.000	.258	.113	.067	.26
Low dilution, high light												
System 1	.500	.079	.441	.060	.008	7.5	.001	0.000	.247	.133	.052	.21
System 2	.505	.089	.416	.056	.007	8.0	.001	0.000	.337	.182	.091	.27
Low dilution, low light												
System 1	.421	.173	.248	.045	.006	7.5	.001	.001	—	—	—	—
System 2	.395	.204	.191	.043	.006	7.2	.002	.001	—	—	—	—

*These values are probably too high since subsequent work has shown that the inorganic medium has an ash-free dry weight greater than its organic (MDTA) content. The ash loss must be due to decomposition of mineral components, but has been interpreted as biomass. Although the relative values can be compared, they should not be considered absolute. This would account for the unusually low C values ranging from 14-22% of the biomass.

†n.d. = not done.

In general, the agreement between replicates was best in the yield reservoirs. This supports the concept that the communities are tending toward a particular stable point and there is more variation on the rates and exact sequence of events leading to that stability than variation at the stable stage.

The concentrations of total biomass[1] were highest under conditions of low dilution and high light (Table 6), but this condition may not have resulted in the highest total production since a smaller volume of culture was produced than during the high dilution period (Table 7). The total biomass production was a function of light, as seen by a comparison of biomass under the two light conditions at low dilution: at half light intensity approximately half the biomass was produced. This result agrees with that obtained by Pipes and Koutsoyannis (1962) on light-limited *Chlorella* cultures. In all cases, at least 84% of the nitrate was converted to organic (Kjeldahl nitrogen (Table 7)). These findings also agree with our findings on light- and NO^-_3-limited *Chlorella:* maximum conversion of NO^-_3 occurs more rapidly than maximum biomass production so that the protein is proportionally greater when total biomass production has been limited.

When the densities of organisms are multiplied by nominal individual weights[2], the biomass can be approximately proportioned. In all cases the algal biomass is at least 85% and sometimes 99% of the total calculated biomass.

Conclusions

The method of gnotobiotic, continuous cultures has been used here with some degree of success for the purpose of gaining insight into community responses to environmental changes. The use of multiple flasks permitted a greater number of conditions to be examined in the same amount of time and with somewhat less than twice the effort that a single stage would have entailed.

The experiment provided documentation that densities, growth rates, and production rates can vary in different directions. This result has long been known on a theoretical basis, but it is still common to assume that lakes with high standing crops are concurrently highly productive.

The algal populations behaved in a highly predictable manner. Had the NO^-_3 and NO^-_2 concentrations been measured in the growth flasks it would have been possible to directly determine kinetic substrate-growth relationships. In the absence of these measurements, it must be assumed that progressive nutrient limitation occurred as the cells traveled down-

TABLE 7.—Total Daily Production (Yield Reservoir Concentration x Volume) of Two Systems Under Three Conditions of Dilution and Light Intensity.

	Biomass[1], mg	Protein[2], mg	% Protein[4] (ash free-dry wt)	Proportion N converted to protein	N recovered,[5] mM
High dilution, high light					
System 1	267.2[3]	27.5	10.3	.84	.506
(green)					
System 2	275.7[3]	30.0	10.9	.90	.527
(red)					
Low dilution, high light					
System 1	172.2	21.0	12.2	1.2	.589
System 2	167.2	16.6	9.9	.94	.475
Low dilution, low light					
System 1	99.5	16.0	16.1	.91	.457
System 2	78.4	15.6	20.0	.87	.439

[1] (See footnote on Table 6.)
[2] Weight N, x 6.25 = weight protein.
[3] No ash weight subtracted.
[4] May be underestimated since biomass may be overestimated.
[5] Greater than 0.5 mM nitrate recovery indicates concentration by evaporation during autoclaving of the medium carboys and during the growth period.

stream. Light limitation was also obviously occurring; in the first flask less mutual shading occurred, but the amount of light received obviously limited production. The further production in the second flask was a function of production in the first.

The inability to determine the factor(s) that limit protozoan density remains a probelm. Certainly availability of algal cells does not provide an adequate explanation. It may be possible to explain some of the discrepancy as a result of the variable size, biomass, and nutritional state of the algal cells. Work has been continuing along these lines, and several apparently obvious hypotheses appear to be disproved, but an adequate explanation has not been found.

The greatest weakness of the method has been its failure to show gross production rates—all of the growth and production rates measured here were determined after consumption of algal and bacterial cells had presumably occurred. The use of ^{14}C uptake as a measure of gross algal production, perhaps in combination with radioautography, may show the amount of carbon fixed by the algae and consumed by the protozoa. Suitable controls could provide an estimate of direct ^{14}C uptake by the protozoa and uptake of labelled extracellular algal products.

Work is still being pursued for calculation of the other transient growth rates, and sample variance is being analyzed for determination of confidence limits and more efficient sampling procedures.

Others are invited to use these data, especially for other modeling purposes. The raw data will be supplied by the author on computer cards with the necessary read-out cards and, if desired, the program used here.

The general method is suggested for use in further experiments on species diversity, including competition, and in pollution experiments with either potentially usable nutritional substrates, such as pulp mill wastes, sugar processing wastes, or sewage effluent, or with sublethal concentration of various toxicants.

Acknowledgments

This investigation was supported by the Federal Water Pollution Control Agency, Grant WP 00982. Contribution No. 337, College of Fisheries, University of Washington, Seattle. I am grateful to Ruth Hung for chemical analyses and culture maintenance, Sheri Lusk Hamel for enumerations and charts, Fred Palmer for NO^-_3, NO^-_2 and pigment analyses, David Drevdahl for computer programming and Jonathan Heller for general help and stimulating discussions.

---

[1]See footnote on Tables 6 and 7.

[2]Alga, 4 x 10^{-8}; protozoan, 1.5 x 10^{-5}, *Pseudomonas*, 2 x 10^{-10}, *E. coli*, 2 x 10^{-10} mg/cell.

Literature Cited

Bendschneider, K. and R. I. Robinson. 1952. A new spectrophotometric determination of nitrite in sea water. J. Mar. Res., 11:87-96.

Difco Manual of dehydrated culture media and reagents for microbiological and clinical laboratory procedures. 1953. Ninth edition. Difco Laboratories, Inc., Detroit, Michigan.

Dougherty, E. C. 1959. Axenic culture of invertebrate metazoa: a goal. Ann. N. Y. Acad. Sci., 77(2): 27-54.

Drake, J. F., J. L. Jost, A. G. Frederickson and H. M. Tsuchiya. 1966. The food chain. *In* Bioregenerative systems, NASA SP-165, pp. 87-95.

Herbert, D. 1964. Multi-stage continuous culture. *In* I. Malek, K. Beran and J. Hospodka (eds.), Continuous cultivation of micro-organisms. Proc. 2nd Symposium held in Prague, June 18-23, 1962. Publ. House of the Czech. Acad. Sci., Prague, 1964, pp. 23-44.

Jannasch, Holger W. 1967. Growth of marine bacteria at limiting concentrations of organic carbon in seawater. Limnol. Oceanog., 12(2): 264-271.

Jannasch, H. W. 1969. Estimations of bacterial growth rates in natural waters. J. Bact., 99(1): 156-160.

King, Elizabeth O., Martha K. Ward and Donald E. Raney. 1954. Two simple media for the demonstration of pyocyanin and fluorescin. J. Lab. & Clin. Med. 44: 301-307.

Monod, J. 1942. Recherches sur la Croissance des Cultures Bacteriennes. Hermann & Cie, Paris. 210 pp.

Novick, Aaron. 1955. Growth of bacteria. *In* Annual review of microbiology, 9: 97-110.

Patten, Bernard C. 1968. Ecological modeling with analog and digital computers. Tutorial lectures and workshops, BioInstrumentation Advisory Council, AIBS, 3900 Wisconsin Ave., N.W., Wash. D.C. 20016.

Pearson, E. A., E. J. Middlebrooks, M. Tunzi, A. Adinarayana, P. H. McGauhey and G. A. Rohlich. 1969. Kinetic assessment of algal growth. Preprint of paper presented at 64th nat'l meeting, American Institute of Chemical Engineers, New Orleans, La. 37 mimeo pages. (Summarized in Joint Industry-Government Task Force on Eutrophication, Provisional Algal Assay Procedure, 1969, P. O. Box 3011, Grand Central Station, New York, N. Y. 10017.

Pipes, W. O. and S. Koutsoyannis. 1962. Light limited growth of *Chlorella* in continuous culture. Appl. Microbiol., 10: 1-5.

Sonneborn, T. M. 1950. Methods in the general biology and genetics of *Paramecium aurelia*. J. Exptl. Zool., 113: 87-147.

Taub, F. B. and A. M. Dollar. 1965. Control of protein level of algae, *Chlorella*. J. Food Sci., 30(2): 359-364.

——————. 1968a. The nutritional inadequacy of *Chlorella* and *Chlamydomonas* as food for *Daphnia pulex*. Limnol. Oceanog., 13(4): 607-617.

——————. 1968b. Improvement of a continuous-culture apparatus for long-term use. Applied Microbiol., 16(2) 232-235.

Taub, F. B. 1969a. A biological model of a freshwater community: A gnotobiotic ecosystem. Limnol. Oceanog., 14(1): 136-142.

—————— 1969b. Gnotobiotic models of freshwater communities. Vehr. Internat. Verein. Limnol. 17: 485-496.

Wood, E. D., F. A. J. Armstrong and F. A. Richards. 1967. Determination of nitrate in sea water by cadmium-copper reduction to nitrite. J. Mar. Biol. Ass. U.K., 47: 23-31.

Wright, R. T. and J. E. Hobbie. 1965. The uptake of organic solutes in lake water. Limnol. Oceanog., 10: 22-28.

Bassett Maguire, Jr.

Community Structure of Protozoans and Algae with Particular Emphasis on Recently Colonized Bodies of Water

Abstract

Twenty beakers, which originally were sterile were filled with artificial lake water and then sampled periodically. Originally sterile polyethylene tubs were placed on the newly formed island, Surtsey. Colonization curves are regular, rising at first rapidly and then progressively more slowly with time. It is possible, however, that the apparent (near) equilibrium will be only "temporary" and that colonization curves are not always monotonic. Correlation coefficient of variation of number of taxa per beaker with mean number of taxa and mean number of total individuals/ml was positive and significant. Diversity [$H' = -\Sigma (n_i/N)(\log_2 n_i/N)$] had essentially zero correlation with coefficient of variation of number of taxa and number of individuals. Autotroph/heterotroph ratio changed through 7 orders of magnitude in the beakers; except in the beginning, autotrophs were dominant. Ecological activity rose with the increase in number of taxa and individuals, but did not decline when the growth rate of either of these slowed considerably.

Introduction

The rates and patterns of colonization and of formation and subsequent change of community structure in initially sterile, really isolated habitats has been studied very little. Such study, especially in very small, isolated aquatic habitats, can be very rewarding because their full biological and physico-chemical history is known, the developing communities are relatively simple, adequate replication and experimental manipulation are possible, and sampling problems are less severe than in other natural environments. Small, isolated habitats, therefore, may be used as especially effective tools for the elucidation of many ecological processes of fundamental importance.

Study of the related processes of colonization and development of community structure on clean surfaces immersed in lakes, streams, or the ocean has produced considerable information. Ruth Patrick has, over a number of years, done some of the best and most interesting of this kind of work; her paper in this symposium reviews some of the results she has obtained through her study of diatoms. John Cairns, also in this volume, writes on what is essentially the same process, although his ecological tools are protozoans in or on sponges.

The first extensive experimental study of the colonization of small, originally sterile bodies of water was carried out in 1959, when it was determined that a surprisingly complete aquatic biota developed in the 265 ml of water in isolated bottles within several weeks (Maguire, 1963a). Raccoons were of major importance in transport of aquatic organisms in this experiment. Colonization rates initially were high and then decreased regularly; number of species per bottle increased rapidly at first and then progressively more slowly toward a maximum determined by the available species pool, the considerably restrictive resources of the small water masses, and the relatively small number of niches provided by a water-filled bottle. A very striking example of interaction was observed in which the ciliate, *Colpoda*, was eliminated from the active community, apparently by biotic mechanims (Maguire, 1963b).

A very stimulating little book by MacArthur and Wilson (1967) gives important theoretical development to our understanding of some aspects of island colonization and community structure, and provides an excellent review and synthesis of recent theoretical work. Some of this theoretical development has been recently tested by experimental defaunation and observation of subsequent recolonization of small mangrove islands of the Florida keys by Wilson and Simberloff and Simberloff and Wilson (1969). Their experimental results generally support the theoretical developments of MacArthur and Wilson.

This paper will consider the rates and patterns of colonization and development of community structure in small bodies of fresh water variously isolated from possible sources of aquatic organisms.

Colonization

Pattern of Increase of Species Number

In late 1963, 33 km off the south coast of Iceland, a volcano arose from the sea to produce the island, Surtsey. Eruptions continued sporadically on Surtsey and on several small islands which formed nearby until the spring of 1967. The last sterilization ash fall occurred in June 1965 (Fridriksson, 1968). Surtsey is about 170 m high and has an area of about 2.3 km². Its southern portion has been covered by flowing lava so that, unless this lava layer is undercut by the sea, the island probably will last for a long time. The other small islands formed near Surtsey were not subsequently covered by lava and as a result were quickly eroded away.

The Icelandic government has realized Surtsey's value to geological and biological scientists, and has placed it under the aegis of the Surtsey

Research Society. This Society, which is composed of scientists, restricts visitation of the island to those who have good scientific reason for going. Those permitted on the island are made aware of the need to keep human transport of plant and animal dissemules to the island at a minimum. The Society publishes a valuable annual report concerning work accomplished on Surtsey (edited by Mr. Steingrimur Hermannsson, the Society's chairman).

In early June, 1967 (as early in the year as it is practical to try to get to Surtsey), 11 aquatic "traps" were placed on Surtsey. They were eliptical polyethylene laundry tubs (80 cm long, 50 cm wide, and 30 cm deep), and they were placed on various parts of the island and held down by lava blocks to prevent them from blowing away. The expectation was that these traps would fill with rain water and that communities of fresh water organisms would then develop in them. Since then the traps have been sampled twice annually, once in early June and again in late August. Aseptic techniques have been used. The traps initially were sterilized by rubbing them with alcohol and all sampling apparatus and jars were autoclaved and kept wrapped or sealed until use. Cultures of sweepings from the cabin floor (where presumably man-imported dissemules would be found in the greatest concentration) have been negative for algae, protozoa, and multicellular organisms.

The summer of 1967 was much drier than usual and the result was that only four of the traps contained any water when the island was visited in the fall. Since then some of the traps have been broken, some filled with blowing ash, one filled with salt water, and one destroyed when the winter storms of 1968-69 eroded about 100 m from the south shore. (These have been replaced or emptied of ash and moved). Because of these difficulties, the sample size is smaller than had been hoped. Nevertheless, when those traps that held adequate amounts of fresh water were considered, the curve produced by the data looks like that I obtained in Texas in 1959 (see below) and is similar to the theoretical curves drawn by MacArthur and Wilson (1967). The average number of taxa per trap grew from 6.75 in the fall of '67 to 15.8 in the spring of '68, 15.4 in the fall of '68, and 16.5 in the spring of '69 (Fig. 1). It might appear that already those traps were approaching equilibrium value. This may be true, and if so shows a remarkably rapid development toward this level. On the other hand, it is possible that the curve will continue to rise for a long time. The species now represented in the traps are probably only the most vagile of the species which live on Iceland and which could survive and grow in the water in the traps. For example, only three or four species of rotifers have reached the traps; all other micrometazoa

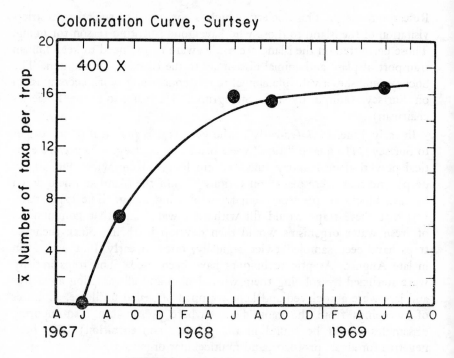

FIGURE 1. Colonization curve: Surtsey.

such as copepods, ostracods, cladocera, gastrotrichs, small annelids, and so on, still are absent. Many algae and protozoans which presumably could live in the traps are also still absent; for example, there are no desmids or peritrichs. It is also possible, as will be discussed below, that species number cannot increase beyond some low maximum until more efficient predators (including herbivores) invade the traps.

Colonization of apparently unfilled niches will proceed at a decreasing rate and will be accompanied by some niche reshuffling and readjustment. The result will be that the number of species will continue to grow (slowly in the absence of predators—see below) for many years, as will the extinction rate, until an equilibrium level is reached. Then there will be a significantly higher number of species than are now observed.

In any event, the detailed shape of this (and any) colonization curve is influenced, sometimes to a considerable degree, by the frequency distribution of vagilities among species of potential colonizers. This point needs greater attention.

The colonization curve for 32 bottles initially filled with sterile water and placed at different distances from a pond and different heights above the ground near San Marcos, Texas, in 1959 is given in Fig. 2. This curve is for the number of taxa that could be distinguished with a dissecting microscope and a magnification of 45X. The circles and the eye fit curve show the colonization rate. Note that this colonization curve (which is on a different time scale from that of Surtsey) is relatively flat. This flatness is in part an artifact of use of data obtained with low magnification; presumably if all species of algae and protozoan were differentiated, the curve would have shown a more rapid rise during the first few weeks, in part the result of different frequency distributions of vagility and availability of protists and of metazoans. The triangles show the relative rates of dispersal and colonization during the periods immediately preceding them. They give evidence that the flatness of the colonization curve is also partly the result of differential input (of the larger species) with respect to time, with the greatest input occurring late in the experiment.

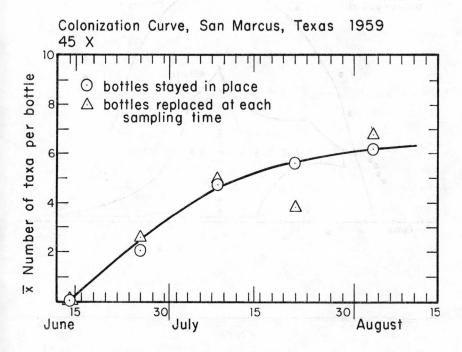

FIGURE 2. Colonization curve: San Marcos, Texas.

There is a good correlation between the rate of colonization during the various inter-sample periods of this experiment and the number of muddy raccoon foot prints on and around the bottles. This, plus comparison with a somewhat similar but raccoon-free colonization experiment to be described later, show that these raccoons provided the dominant fraction of the input to these bottles. See Maguire, 1963a for further details.

In the spring and summer of 1969 a number of hard working students and I carried out another colonization experiment in which we gathered information in an attempt to show, in some detail, the changing community structure throughout the early phases of colonization.

On March 12, 1969, 10 beakers of 800 ml capacity and 10 beakers of 400 ml capacity were placed 3 meters apart in a cross shaped pattern in a grassy field near the Colorado River in Austin, Texas. Fig. 3 shows their arrangement and the location of trees, a wire fence, and the river. An electric shock fence (not shown on the map) kept raccoons and other larger animals from affecting the experiment. Note that the beakers in

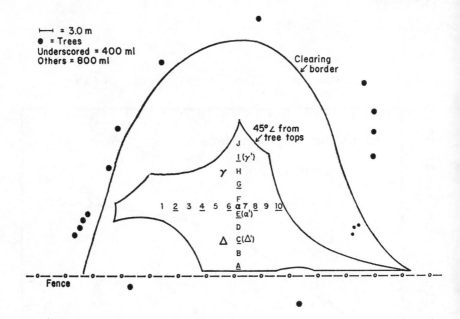

FIGURE 3. Map of Austin colonization experiment.

line perpendicular to the river are lettered consecutively; A is closest to and J farthest from the river. The central-most (scalloped) line indicates where in the field the tops of the nearest, and/or tallest trees were at 45° altitudes. Large and small beakers were alternated in the "cross"; the small ones were suspended with stainless steel wire harnesses in 800 ml distilled water-filled "overbeakers," so that the total volume and surface/volume relationships of the 400 and 800 ml systems were the same. Their temperatures, therefore, should have been the same. All beakers were fastened to steel posts, the tops of all beakers were 60 cm above the ground and all were shaded by a 25 x 25 cm sheet of galvanized iron held 10 cm above them. Each shade held a funnel which emptied into the beaker below and had the same diameter as the beaker into which it emptied. Beaker and funnel diameters were 75 and 100 mm for small and large respectively. This produced rain collecting and air/water surfaces which were the same per unit volume for large and small beakers.

All beakers, funnels, and shades were sterilized, and the beakers were filled with artificial lake water (see Maguire, 1963a). Collections were made at 2-day intervals for almost a month (when analysis time increased to the prohibitive point), then at 4-day intervals for another month, and later at irregular intervals as time permitted.

Aseptic techniques were used. At each sample time the beakers were first filled with sterile, distilled water (to compensate for the evaporation which had occurred), stirred very vigorously, and then sampled with effort to take parts of the sample from each part of the beaker. Sampled volume was replaced with sterile artificial lake water. (During the last two months some collections were made prior to filling of the beakers.) Samples were kept cool until they were examined, always on the same day.

Initially, when organism densities were somewhat low, species' estimates were made from counts from five transects of a Sedgwick Rafter cell using an ocular micrometer and a magnification of 100. When population densities increased to the point that counts became very high, we changed to a chamber made by supporting a No. 1½ cover slip on flanking No. 1 coverslips (thickness 0.14-0.17 mm, assumed to be 0.15 mm). The coverslips were held down by capillary forces, as care was taken never to completely fill the chamber. (Occasionally small metal weights were used—apparently without effect). Transect length was 15 mm and indicated by two lines scratched on a microscope slide parallel to its edges. Five transects were counted. Filamentous algae were counted only when they were intersected by a given line of the ocular micrometer and filament length/ml is calculated: $\text{mm/ml} = \dfrac{\text{(total intersections for 5 tranment)}}{5}$

sects) $\left(\dfrac{1}{\cos 45°}\right) \left(\dfrac{1000}{(15)\,(.15)}\right)$. Results of change in counting technique are not detectable in the data.

As filamentous, sheath-producing algae (esp. Oscillatorales) became common, it was necessary to count them, and many of the Chlorococcales and other small algae which became imbedded in or stuck to their gelatinous products, after blending the sample for 30 seconds in a micro-cup attachment of a Waring Blendor. Fragile algal species and most protozoa could not withstand this treatment and so were counted before the sample was blended (experiments demonstrated that the number of the more robust algae was not affected).

Effect of Distance from the River

Analysis by the rank difference correlation coefficient, r_d, (Tate and Clelland, 1957) of each sample series through May 19 did not demonstrate a significant difference in number of recognized taxa as a function of distance from the river. Average r_d from March 24 through May 5 was 0.24 [r_d is significant ($n=10, p=.05$) at .65]. On May 11, r_d was .66, on May 15 it was .84, and then on May 19 it had dropped, without obvious cause, to $-.12$. On the whole, there is a suggestion that the river may have preferentially contributed a few more taxa to the beakers near to it, but the effect if it exists is small.

Comparisons of Large and Small Beakers

The average number of taxa observed (excluding bacteria and fungi) per 800 ml beaker (from March 24 through June 23) was 8.87 and the mean for the 400 ml beakers was 8.51. If the period of early colonization is excluded by choosing the interval April 9 through July 19, the means are 9.39 taxa for the large beakers and 9.55 for the small ones. These figures suggest that there is no difference between number of taxa as a function of beaker size when sample size totals 20 and size difference between beakers is 1:2. Perhaps, however, colonization rate of the larger beakers may have exceeded (slightly) that of the smaller ones during the initial phases.

The average number of organisms (excluding bacteria and fungi) per ml between April 9 and July 19 was 3.0×10^6 for the large and 2.4×10^6 for the small beakers, but the difference does not approach statistical significance. (The April 9-July 19 period was one in which increase in number of individuals was approximately exponential, but at a rate considerably less than that which occurred as the first colonists of the beaker underwent the initial population "explosion" during which they "filled" the habitat—see below.)

Input Pattern

Three beaker stands (α, Δ, γ) similar to those described above were used to measure dissemule input to the experimental beakers as a function of time. Figure 3 shows their location with respect to the experimental stands. Sterile, water-filled beakers were placed on these stands at each time of collection and the water which had been sterile at the preceding sample time and therefore contained 2- or 4-day input was used to provide innoculum for flasks of Bold's Basal Medium (Bold, 1942), which is designed to provide nutrients necessary for the growth of many species of algae. These were cultured under a mixture of fluorescent and tungsten light for daily 12-hr. periods, and examined at periods of several weeks. These data gave important information on the rate of input of dissemules into the beakers, showing that it was somewhat irregular, but that there was no discernible pattern through time except for the possibility that input may have been slightly greater during the first months of the experiment when rain fell most frequently. Of the algae which grew from cultures taken from March 16 through May 3, *Chlorella* had by far the highest vagility and/or availability, as it was found in 16 of these cultures. *Chlorococcum* was second with seven occurrences, *Oocystis* and Bodonids had three, and *Palmellococcus*, Diatoms, *Oscillatoria*, *Kirchneriella*, *Rhizochlonium*, *Scytonema*, and *Coccochloris* were seen once or twice.

Colonization Curve

Figure 4 illustrates colonization pattern in terms of average number of taxa observed per beaker as a function of time. The theoretically expected colonization curve is approximated by the line. This expected colonization curve fits points after April 4 nicely, but there is considerable deviation between the observed pattern and the expected pattern at the beginning of the experiment. This is the result of the methods used in making the estimates of the number of taxa.

Knowledge of the time of each successful colonization is necessary to demonstrate the classical colonization curve. During the early parts of the beaker study this curve shape, which most probably followed the

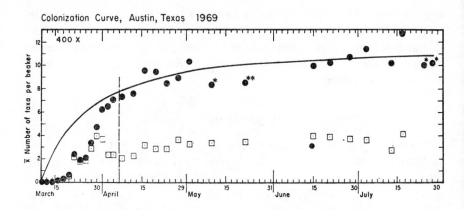

Colonization Curve, Austin, Texas 1969

FIGURE 4. Colonization curve: Austin, Texas (this study).

expected form, was obscured. This was because, after different numbers of dissemules of the various species arrived in the beakers at different times, they then required different times for excystment, and reproduced at different rates. Therefore, different amounts of time elapsed (after arrival) before they reached densities at which they could be observed. Each variable (number of dissemules, time of arrival, excystment time and reproductive rate) will tend to have a normal distribution; cumulative normal curves are of sigmoid shape.

Although in some ways it is a disadvantage that there must be at least several cycles of reproduction of a species in the new habitat before that species will, on the average, be detected, it may turn out that this disadvantage will be outweighed by the advantage resulting from better representation of the biological realities of colonization than otherwise would be possible. The definition of colonization as being the arrival in the new area of a single potential reproductive individual or pair (or more) is convenient and lends itself well to theoretical manipulation as it has been carried out by MacArthur and Wilson, and by Wilson and Simberloff. In biological terms, however, a definition of colonization as the arrival and subsequent reproduction (through several generations?) of a new species may be more useful. The problem, which the definition of MacArthur and Wilson avoided, is that it is difficult to decide how much reproduction is enough to qualify the new species as a colonist. In some species it might reasonably be more than several generations. For example, it is well known that some species of algae require vitamin B_{12}

and can store enough to permit growth and reproduction through several generations. Should one of these B_{12} requiring cells obtain a large amount of the vitamin and then be transported to an environment devoid of B_{12}, all of the cells will die, but only after several generations of reproduction. It seems to me that this species could only be a considered transient or a noncolonist (failure as a colonist) but never a colonizer of this B_{12}-free habitat. Colonization as a biologically meaningful activity can only occur when the species potentially can survive the conditions of the new environment for many generations. Under these conditions a colonizing species can also be thought of as having an appreciable ecological effect on the environment and/or the populations already living in the area. The difficulty in delimitation of the proposed definition should be more than compensated for by the more realistic biological meaning that the term would have.

As far as one can tell from an examination of the data, the number of taxa colonization curve for the Austin 1969 experiment is continuing to climb slowly. It may be that this will continue for a long time for reasons outlined above.

The difference between the increase in number of taxa of heterotrophs (circles) and that of the autotrophs (squares) as seen in Fig. 4 is interesting. The heterotrophs made up nearly all of the total during the early part of the colonization (until March 28); then after March 30 the curves separated rapidly and the number of kinds of heterotrophs remained at about three to four per beaker, while the number of taxa of autotrophs continued to increase to about 11. The heterotrophs were primarily Monads, Bodonids, a few small Holotrichs, and a few small Amoebae. Before the growth of algal populations they apparently fed on bacteria growing on the pollen grains, leaf bud scales, anthers, and small unidentifiable pieces of organic debris, which were fairly common almost immediately after the beakers were put in the field. The autotroph/heterotroph ratio will be discussed at greater length below.

Equilibrium from Above

In an attempt to determine equilibrium level (see MacArthur and Wilson for development of this concept) for these beakers in a different way, a large number of taxa were added to four of the beakers on July 20. These were obtained by collecting samples (somewhat concentrated with a plankton net) from two rivers, a permanent pond, and water in which mud from the bottom of some ephemeral ponds had been cultured for several days. These samples were then stirred together and the resultant taxon-rich mix was used to replace about 50 ml of beakers 1, 4, 6, and 7.

These beakers therefore contained the communities which had developed in them plus a large number of new taxa. The number of taxa which had made up the communities in these beakers was about 12 on the average; on July 23, three days after the mix was added, the beakers contained an average of 25.5 taxa. This number did not change appreciably by September 2, the last day that they were sampled (when they also contained 25.5 taxa, although there had been a little fluctuation between these dates).

One might theorize that an input of a large number of taxa into beakers in which colonization had proceeded for four months would be followed by violent ecological activity resulting in rapid modification of the structure of the community that had developed through the operation of natural mechanisms of dispersal, colonization, and community development. This activity presumably would quickly result in some reduction in the number of taxa present. With time as the more long term extinction mechanisms operate, the number of taxa will gradually decrease and approach the equilibrium value from above.

The lack of change in the number of taxa from July 23 to September 2 suggests either that 25 is at or below equilibrium number for these beakers or that extinction rates had no effect in six weeks.

It would not have been surprising if the addition of new taxa, including several kinds of micro-crustacea, would have resulted in a rapid reduction of number of taxa. I observed natural events of this kind in rain water caught and held by the flower bracts of the wild banana, *Heliconia*, in the forest at El Verde Experiment Station, Luquillo Natural Forest, Puerto Rico. Protozoan communities develop very high densities in the water of some of these flower bracts, and then when mosquito larvae appear, their feeding activities are so efficient that the protozoan populations are reduced to very low densities, frequently with extinction of many of the taxa which had been abundant (Maguire, 1968).

It is possible that, given enough time, the Ostracods and Copepods added to the beaker communities will have a similar (though less extreme) effect to that of mosquito larvae on *Heliconia* communities. On the other hand there is a very interesting alternative hypothesis which may have great implications with respect to both theoretical and practical considerations of island colonization. It is possible that the equilibrium value for the beakers is 25 or more taxa, but that this cannot be even approached until one or more relatively efficient (but not too efficient) herbivores which feed on the originally dominant alga taxa, are present. It seems probable that a community in a beaker would frequently act to prevent colonization by new immigrants through mechanisms discussed by Elton in his *The Ecology of Invasions by Animals and Plants* (1958). This is especially probable where, for example, several species of algae are greatly dominant because they are most rapid and efficient in utilization of algal nutrients that become available. A rather simple algal community could be indefinitely stable under these conditions. The communities which developed in the beakers that relied only on natural mechanisms for dissemule input may have reached close to an equilibrium of this kind. (The protozoa and rotifers which colonized the beakers consumed some algae but were observed to have eaten algal cells sufficiently infrequently to suggest that they were not effective in limiting the algal populations.) This kind of intermediate equilibrium could be maintained for a long time, and might well include a much lower number of species than could exist in the beakers (or on islands) at highest equilibrium.

A major discontinuity in the colonization curve could then be produced by the colonization of the beaker (or island) by one or more efficient herbivore(s). The results from the addition of many taxa to the beakers as described above suggests that the additional number of taxa that might be permitted by the activities of the herbivores could be as great as the number at intermediate equilibrium. It is conceivable that a second and higher intermediate level might occasionally occur in the absence of carnivores. Then their addition would be necessary before highest equilibrium could develop. In either event the colonization curve is not necessarily monotonic. Rather it could be of the shape shown in Fig. 5. In this curve the dotted line illustrates the average course of colonization when efficient herbivores and carnivores enter the community appropriately soon after the beginning of the colonization process. The solid line illustrates the expected average (idealized) course of colonization of beakers (and islands) where efficient herbivores have significantly poorer dispersal mechanisms and therefore considerably later times of arrival than the autotrophs, and the carnivores have lower dispersal rates than the herbivores, and therefore arrive still later.

This colonization pattern is consistent with the findings of Paine (1966) in which the number of species on intertidal rocks was considerably increased by the presence of a moderately efficient predator.

If, however, the predation (including herbivores feeding on autotrophs) is exceedingly efficient and involves a large enough number of prey species, a "final" equilibrium lower than the intermediate equilibrium (and much lower than highest equilibrium) can be established. The effect of mosquito larvae on communities held by *Heliconia* bracts (given above) provides a good example.

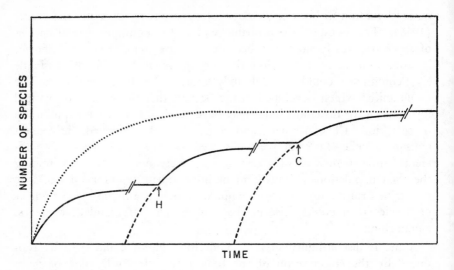

FIGURE 5. Non-monotonic colonization curve. H and C indicate time of arrival of adequately efficient herbivores and carnivores respectively. See text for other details.

In sum, it is the interaction between the different colonizing potentials of some groups of the herbivore and carnivore trophic levels, and the competition-reducing effects of these groups on the trophic level below them, that can produce immigration, extinction, and colonization curves which are not monotonic.

Change in Number of Individuals

Fig. 6 illustrates the average number of individuals/ml of all taxa as a function of time. The number increases at a rapid rate, which is approximately exponential for about the first month. During this time it passes through six orders of magnitude. Then it abruptly changes to a smaller slope but continues to increase exponentially. Two straight lines, both passing through the April 8 data point, could be drawn and would closely fit all points. This April 8th point also could be the one that might be chosen if it were desirable to pick a point representing the ending of the most rapidly rising part of the taxa colonization curve. It would appear that the high initial rate was the result of rapid growth during the "filling" of the environment under conditions of little or no competition. The later, slower growth probably was accompanied by considerable re-shuffling of resource utilization through differential growth of the various

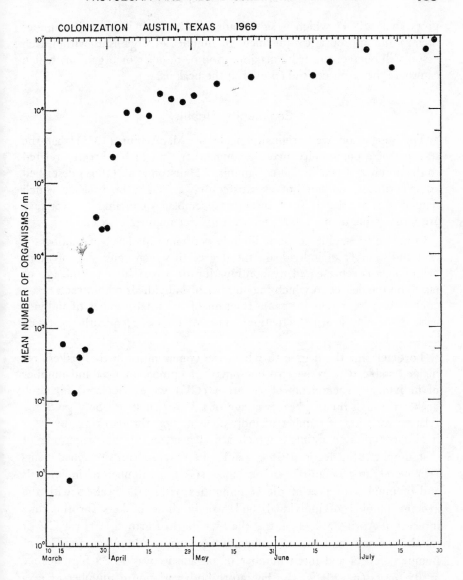

FIGURE 6. Mean number of organisms per beaker through time.

taxa. This reshuffling would, other things being equal, provide for more efficient use of the total resources and permit increase in number of organisms (and number of taxa).

Comparison of the mean number of taxa per beaker and the mean number of individual organisms per ml during the April 9-July 19 period gives an r_d of .47 which is significant at .05. Both increase in number of taxa and appropriate adjustment of number of organisms within each taxon will operate toward a maximization of number of organisms which can share the resources and co-exist in the beakers.

Community Stability

The suggestion was originally made by MacArthur (1955) that the stability of a community may (among other things) be directly related to the number of taxa in that community. Hairston et al. (1968) described an experiment, though not as satisfactory as might be wished, which tended to show that in some instances assemblages containing fewer taxa are more stable than assemblages with greater number of taxa.

Community stability criteria include constancy of level of number of taxa and number of individuals, the degree to which constancy of number of taxa is represented by continuation of a particular assemblage of taxa, and the degree to which the number of individuals within each taxon continues to represent the same fraction of the total number of individuals. Normal seasonal fluctuations must, of course, frequently be taken into account.

To determine the degree to which the communities in the beakers remained stable with respect to constancy of number of taxa and number of individuals, coefficients of variation (CV) were calculated for each beaker through time. The advantage of CV is that it not only gives the relative variation of number of individuals or taxa through time, but that it also provides an estimate which is independent of the magnitude of that number (Snedecor, 1956). Rank difference correlation coefficients (r_d) were run to compare the stability (CV) of number of individuals and of number of taxa of the communities within the beakers with the average number of individuals and taxa in those beakers for the time interval of April 9-July 19 (after the first spurt of growth): r_d for CV of number of taxa and mean number of taxa was −.43, and r_d for CV of number of taxa and mean number of organisms was −.53. Significance at .05 for r_d (n=20) is .45. Therefore both variation of number of taxa and variation of number of organisms in a beaker appear to have been reduced by the presence of greater number of taxa. In other words, in these systems (and at this stage of succession—community development), stability increases with increase in number of taxa present. CV of number of organisms, however, did not show significant correlation coefficients with average number of organisms, average number of taxa, or CV of number of taxa (r_d was .30, .26, and .12 respectively).

Diversity

Diversities, estimated by $H' = -\Sigma - \frac{ni}{N} \log_2 \frac{ni}{N}$, were computed for each sample. The means of the diversities for all beakers, are plotted as a function of time in Fig. 7. These increased during the early period of rapid colonization in a form very similar to the change of number of taxa curve. The diversities, however, increased even more rapidly than the number of taxa and peaked about a week before the taxa data points caught up with the expected colonization curve. This peak was also a week earlier than the break in the number-of-individuals curve. The diversity curve then may have undergone something of a dampened oscillation before it settled down to a rather constant value.

The initial high peak is primarily the result of simultaneous growth of populations of several species to somewhat comparable high levels. Then, some of the species were able to continue their population growth and reach very high densities, while other species, presumably at least partly because of reduction of resources (nutrient?) to below their growth requirement level, were unable to maintain their rate of increase. This produced a community in which one species became greatly dominant, with resultant reduction of diversity.

Fig. 8 shows the change of number of the most numerically dominant taxa (which also make up the bulk of the biovolume) for beaker F, which had a pattern of diversity change through time different from that of the majority of the beakers. In this developing community, diversity increased very rapidly between March 26 and March 28; on the latter date it was higher than for any but one other beaker of the whole experimental series. The reasons for this were that four taxa (two ciliates, a monad and a bodonid), an unusually high number for so early a date,

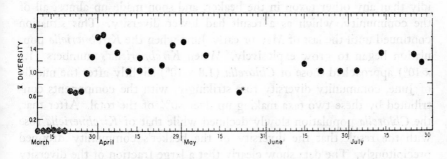

FIGURE 7. Mean diversity of beaker communities through time.

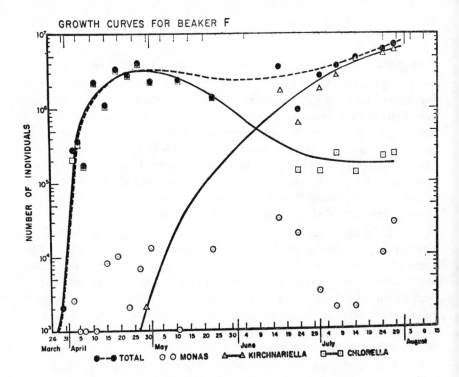

FIGURE 8. Beaker F species curves.

were present and represented by approximately equal numbers of individuals. This equality of number of individuals, of course, resulted in a H' which was nearly as high as it could have been (near H_{max}) considering the number of taxa present.

Immediately after this, however, *Chlorella* increased much more rapidly than any other taxon in the beaker, and soon made up almost all of the community, which as a result had lower diversity. This situation continued until the last of May or early June when the *Kirchneriella* population began to grow explosively. When *Kirchneriella's* numbers (1.7 x 10⁶) approached those of *Chlorella* (1.8 x 10⁶) slightly after the middle of June, community diversity rose strikingly, with the components contributed by these two taxa making up over 80% of the total. After that, the *Chlorella* population slowly declined while that of *Kirchneriella* rose with the result that the diversity of the beaker's community dropped precipitously. The data show clearly that a large fraction of the diversity (H') of the community in the beakers frequently results from equitability of two or three abundant species, and then may show negligible response to the number of taxa present.

Oscillatoria began to become important in most beakers toward the end of June or early July; beaker F was no exception. The dominance with respect to volume by the middle of August was held indisputably by *Oscillatoria*, although numbers of *Chlorella* and/or *Kirchneriella* frequently were also high.

If one of the ecological results of a high diversity is stability, then it seems reasonable that there would be an inverse correlation between the diversity and the amount of variation of the community through time. The coefficient of variation (CV) of number of organisms/ml in each beaker through time was compared with the mean of that beaker's diversity by the rank difference correlation coefficient; r_d was —0.05. This correlation is so low that it fails to support any relationship. If mean diversity is compared with CV of the number of taxa, r_d is 0.18 which also is too low to be meaningful. This suggests that if either diversity or coefficient of variation is a good measure of stability the other is not. Since the coefficient of variation as used here is a good and direct measure of degree of fluctuation of numbers of organisms and number of taxa through time, evidence is good that the relationship between diversity and stability is not close. This is because of the confounding effect of the equitability and number of taxa components of the diversity index. It suggests that difficulties may follow an attempt to use a simple numerical index which summarizes (and confounds) data of dissimilar kinds.

Several less interesting relationships were suggested by the data. When mean total biomass of all organisms and mean diversity are compared, r_d is —.28; when the comparison is between mean total number of organisms and mean diversity, r_d is —.44, and when it is between mean number of taxa and mean diversity, r_d is .21. An r_d of .45 is significant at the traditional .05 level (N=20). The only one of these r_d's likely to be meaningful is that between the mean number of organisms and mean diversity, and this is negative, which suggests that those beakers which, on the average, had the greatest number of organisms also had the lowest diversities. This resulted from the tendency for a single species, usually *Chlorella*, to completely dominate the community (both in number of individuals and in biomass).

Autotroph—Heterotroph Ratio

Fig. 9 shows the change in the autotroph to heterotroph ratio in the beakers from the middle of March to the end of July. The early part of the curves are not surprising because they are consistent with the early invasion by the small heterotrophs followed by the invasion of autotrophs

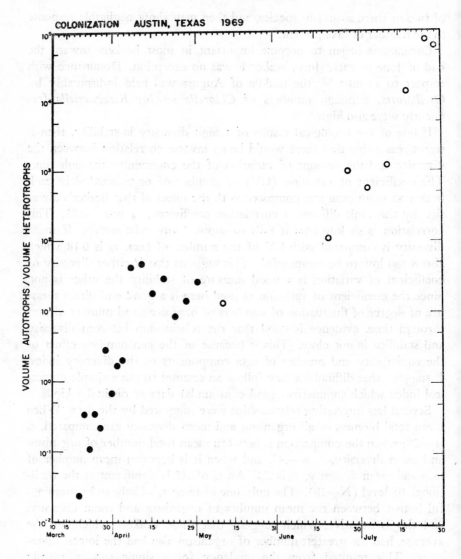

FIGURE 9. Auto/heterotroph ratio through time.

which was illustrated in the taxa colonization curve of Fig. 4. The surprise was that the ratio then continued to rise, and went so far (through over seven orders of magnitude—from about 10^{-2} to over 10^5), the com-

bined result of the growth of the populations of algal taxa and a diminution in the densities of the heterotrophs. Most of the heterotrophs apparently were bacterial feeders, although some were seen to ingest algal
cells.

If heterotrophs capable of ingesting the algae reached the beakers, and
it appears that some did, then there are several alternatives to consider.
Perhaps only some algal taxa were edible and they were soon eliminated.
Alternatively, at least several of the algal species may have released substances which very effectively inhibited the heterotrophic populations.
Or the algae (and the bacterial populations which they support), at least
in beaker communities which have been developing for three or more
months, do not contain enough usable food (or sufficient localized concentrations of food) to provide for the growth of those heterotrophs
which have been able to invade the water in the beakers. It may be that
critical changes occurred in chemical content of many algal taxa as the
populations aged, and, as Fogg's (1965) data suggest, nutritional value
declined. There was a general change in color of the total algal population from green to yellow or yellow-brown, during the study. This kind
of color change coincides with algal nutrient starvation, which is accompanied by reduction in nutritional value of the algae (for example, there
is a lowered protein content).

Chlorella is known to produce an antibiotic which has inhibitory effects
on other species (Fogg, 1965), and Silvey, in this symposium has discussed
these kinds of interactions. *Oscillatoria* also is known to have inhibitory
effects, and its biomass began to become appreciable by mid to late May.

Even though the effects of inhibitory metabolites may be great, and
could explain the great increase in the autotroph-heterotroph ratio which
has been observed, it still may be useful to examine another possible major
explanation. Ecologists frequently assume that whenever food appears
(to them) to be present there must be plenty that the heterotrophs can
eat. Hairston, Smith and Slobodkin (1960), and Slobodkin, Smith, and
Hairston (1967) have suggested that all trophic levels except that of the
herbivores are, as a whole, food-limited. The argument is that since the
world is green, obviously there is plenty of food for the herbivore trophic
level, and therefore that food limitation cannot control this trophic level.

The major difficulty with this hypothesis is that it considers food quantity and does not take quality into account. Required substances may be
absent or may be present but not available. In Slobodkin, Smith and
Hairston's 1967 paper, a table shows that various species, of different
trophic levels, utilize different proportions of the net production of the
food which is assumed to be available to them. The herbivores have much

lower percent utilization rates than do members of other trophic levels which they list.

Even if available energy content is adequate, other necessary nutrients may not be available or sufficiently abundant or concentrated. For example, both vitamin and protein levels may be inadequate; they differ from one part of a plant to another, from one plant to another, or within a single plant at different times. In addition, plant physical characteristics change with time; for example, cell walls become heavier, with the result that breaking open the wall to get to the cell contents (which may be reduced) becomes more difficult. Anyone with a flower garden knows that the small herbivores which attack plants frequently preferentially eat the young growing plant parts. Of the vegetative parts, they have the highest protein and water content and the thinnest cell walls. Flower and fruit, which also are frequently eaten preferentially, have especially high concentrations of easily digestable carbohydrate, protein, and other valuable nutrients. Perhaps the reason that the world is green is that herbivores, as a group, have difficulty in finding enough food with high enough nutritional quality to enable them to survive, grow, and produce strong offspring. Population control by food limitation must not be thought of as operating only through the production of animals which are obviously starving. It is more likely to operate most of the time through reduction in vigor, producing an increase in mortality and decrease in reproductive rate. These changes in mortality and reproductive rates do not need to be large to be very effective.

Taub, in this symposium (and 1969), outlines some of the problems related to constructing simple self-sustaining gnotobiotic communities. One major problem, as her valuable and persistent work demonstrates, is providing the top heterotrophs (protozoa) with food. They need organisms which they will eat and which have sufficient, usable nutrient of adequate quality to permit them to reproduce enough to at least maintain their population. Taub and Dollar (1968) make this point even more forcefully in their paper describing the nutritional inadequacy of *Chlorella* and *Chlamydomonas* for *Daphnia*.

An accurate summary of trophic level limitation is: All trophic levels are, in general, food-limited. I consider this point in greater detail elsewhere (Maguire, in ms).

Interspecific Association Patterns

In the study of colonization of bottles in San Marcos, Texas, when raccoons were active in the transport of a large number of organisms from the pond to the bottles, there were a large number of positive association coefficients (Cole, 1949) between the species involved. These association coefficients reflected common occurrence of the species in the pond and their simultaneous transport to some bottles and not to other bottles. The only striking exception to this pattern was that of the ciliate, *Colpoda*, which was transported to many of the bottles by rapid, non-raccoon mechanisms. Shortly after the assemblage of species carried by a raccoon was added to a bottle which held active *Colpoda*, the ciliate either encysted or was eliminated (Maguire 1963 a+b).

The association patterns between the species which have colonized the beakers in the Austin study is very different. They were calculated for March 28, April 3, July 2 & 7, and July 14 & 18 communities. Out of the 184 association coefficients calculated for the taxa which occurred with sufficient frequency (and not with too great a frequency), only four were "significant" at the .05 level. On the basis of chance alone, one would expect to find nine "significant" associations in a sample of this size. This is evidence that there was, at most, little very strong interaction between taxa which were intermediate in frequency of occurrence. This does not mean, however, and this point should be emphasized, that there is evidence for lack of any interaction. It is very possible that there is much interaction which is not mandatory to some species' occurrence in a given beaker or not so strong that it produced the elimination (exclusion) of species from the beakers. It also is very possible that there is interaction, either positive or negative, but that it occurs between two taxa, at least one of which has great enough vagility and colonizing ability to develop populations in at least most of the beakers. This species could easily have had a major effect on other species, and the association coefficients would be either very high or very low, but they would not be meaningful because of the extreme breadth of the confidence band. Another kind of experiment will be necessary to adequately define these interactions.

Small, isolated bodies of water do not always have the very low level of associations measurable by use of the beaker data. Fig. 10 shows part of the community structure of organisms living in rain water caught in the leaf axils of the Bromeliad, *Guzmania*, in the forest of El Verde Experiment Station, Puerto Rico. The solid lines indicate positive association, and the broken lines show negative association. The thickness of the line indicates the general magnitude of the association. (The thinnest lines show associations which are relatively large, but a little below significance level of .05). In addition to illustrating the sign and magnitude of the strands of the web of interaction between these species, this diagram also tells us that, at least when these data were collected, Cyclopoid copepods were having much greater reproductive success than were Harpac-

Association Pattern in <u>GUZMANIA</u> Communities

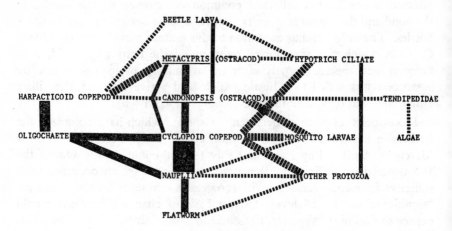

FIGURE 10. *Guzmania* **community structure.**

ticoids. There was, for that matter, a negative coefficient between Harpacticoids and Nauplii, but it was small enough to have little meaning in itself. For more detail see Maguire, in press.

Ecological Activity

The effect that each species has on the community is a function of the amount and kind of resources acquired, held, and released per unit time. In a stable community, the long term rate of acquisition of resources will be balanced by the rate at which they are released; estimates of rates of uptake or release give measures of one kind of a species activity. The amount of each resource stored in or occupied by a species is also of importance because it represents resource which either cannot be used by a competitor and/or is held in a form which is more or less available to predators (including herbivores).

Rates of production or of metabolism were not measured directly in this study; the only pertinent data available concern changes in the number of individuals and bio-volume of each taxon. Obviously considerable ecological activity can take place when the number of organisms within each species is constant, not only through simple turnover of matter and energy by the species, but also through change in the amounts and kinds of materials making up the individuals in the species populations. Fogg (1965), for example, reports on some of the kinds of changes which occur in the size and chemical composition of algal cells as a species population ages. Decrease in photosynthetic and metabolic rates also can occur in aging (nutrient-starved) algal populations.

A more obvious ecological activity is expressed by change of community biomass and/or by a change in the relative fraction of the community represented by each species. In the beakers there was a rapid initial exponential increase in biovolume (biomass), followed by a continued but much slower exponential rate of increase. The solid circles of Fig. 11 illustrate this pattern, a pattern which expectedly follows that of number of organisms (Fig. 6). Presumably what happened was that there was a rapid initial "beaker-filling" spurt of growth which was followed by slower but continued increase as interspecific adjustments occurred which resulted in a rather slow but continuous (through the time of this experiment) increase in efficiency of resource utilization (and therefore of biovolume) by the community. The open circles and triangles of Fig. 11 illustrate the average summed change in biovolume of all species within the beaker communities (increase and decrease of each species biovolume were added without regard to sign). The striking feature of these curves is that the rate of change remained so high after April 9 when the initial beaker-filling growth spurt was over. This is evidence for intense ecological activity during which biomass was, in effect, exchanged between species. These kinds of activity presumably lead toward a more stable community in the long run; in the short term duration of this experiment they certainly seem to lead to more efficient utilization of the beaker's resources.

Summary

1. Colonization experiments with initially sterile water on the newly formed volcanic island, Surtsey, and in bottles in Austin, Texas, demonstrate, with expected deviations, normal and expected colonization curves. Species number increase rapidly at first and then with decreasing rate through time.

2. In both instances there are reasons to suspect, although species number had apparently reached a nearly constant level, that the apparent species equilibrium number was considerably lower than the final value eventually will be. The "temporary" intermediate equilibrium may, however, last for considerable time.

3. Evidence is available to suggest that immigration, extinction, and colonization curves may not always be monotonic.

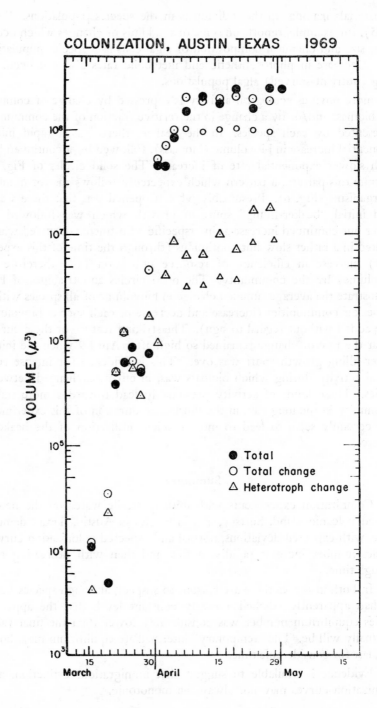

COLONIZATION, AUSTIN TEXAS 1969

4. In some systems, colonizing potentials of efficient herbivores and predators are considerably lower than those of the autotrophs. This, in conjunction with the competition-reducing effects of grazing and predation, explains the lack of monotonicity of the curves listed above.

5. In the Austin study:

(a) Distance from the river had little influence on number of taxa that became established in a beaker.

(b) No difference in number of taxa was observed as a function of beaker volume (400 and 800 ml).

(c) Input pattern did not vary appreciably with time.

(d) Mean number of taxa per beaker through time was positively and significantly correlated with mean number of total individuals/ml.

(e) Coefficient of variation of number of taxa per beaker through time was negatively and significantly correlated with both mean number of taxa and mean number of total individuals.

(f) Diversity ($H' = - \Sigma \frac{n_i}{N} \log_2 \frac{n_i)}{N}$ was computed, and its correlation with the coefficient of variation (CV) of number of taxa and CV of number of individuals m/l was very close to zero. The confounding effects of number of taxa and the equitability component of this index render it useless as a measure of stability under the conditions of this experiment.

(g) The autotroph/heterotroph ratio (both number of individuals and biovolume) changed through over seven orders of magnitude, with the autotrophs always, except at the beginning, being greatly dominant.

(h) This development of autotroph/heterotroph ratio could have been the result of antibiotic production by some of the algae, but more probably was because the algae became progressively less nutritious and less edible. Lack of availability of enough food of adequate quality was probably limiting to the protozoan populations, and to all trophic levels in general.

(i) Fewer interspecific associations were observed (the measuring method was very crude) than would have been expected by chance alone (the difference was much too small to be significant), although this is not always the pattern in similarly small natural bodies of water.

FIGURE 11. Mean total biovolume per beaker, and rate of change (per 2 days) through time. Rate of change is ½ the difference in volume between time t and t+1 plus ½ that between t and t—1, summed for all species. The small figures represent change rates which occurred if division of the measured change by 2 was appropriate after inter-sample periods were lengthened from 2 to 4 days. (This procedure is appropriate only to the degree that the average species population fluctuated with a "period" of 8 days or longer.)

(j) Ecological activity rose with the increase in number of individuals in the beakers. When total growth rate decreased, the ecological activity did not fall off, but rather remained high and was expressed by a re-shuffling of species as, in effect, biovolume was transferred from one to another.

Acknowledgments

Thanks go to the National Science Foundation, the Surtsey Research Society, and the University Research Institute of The University of Texas at Austin for financial support of parts of this study. I wish to express appreciation to Mr. Steingrimur Hermannsson, Chairman of the Surtsey Research Society for his help with logistics of the Surtsey study. I also am grateful to the following students whose work made the Austin coloni-zation study possible: Sue Anderson, Bill Gorman, Bill Houk, Steve James, Claudia Jones, Russell Mase, Jay Wilson, Dan Udovic, and Ed Zyzner.

Literature Cited

Bold, H. C. 1942. The cultivation of algae. Botanical Rev. 8:69-138.

Cairns, John, Jr. 1971. Factors affecting the number of species in fresh-water protozoan communities. This symposium.

Cole, L. C. 1949. The measurement of interspecific association. Ecology 30:411-424.

Elton, C. S. 1958. The ecology of invasions by animals and plants. Methuen. London. 181 p.

Fogg, G. E. 1965. Algal cultures and phytoplankton ecology. University of Wisconsin Press, Madison. 126 p.

Fridriksson, S. 1968. Life arrives on Surtsey. New Scientist 684-687.

Hairston, N. G. et al. 1968. The relationship between species diversity and stability: an experimental approach with protozoa and bacteria. Ecology 49:1091-1101.

Hairston, N. G., F. E. Smith, and L. B. Slobodkin. 1960. Community structure, population control, and competition. Amer. Naturalist. 94:421-425.

MacArthur, R. H. 1955. Fluctuations of animal populations and a measure of community stability. Ecology 36:533-536.

MacArthur, R. H. and E. O. Wilson. 1967. The theory of island biogeography. Princeton University Press. 203 p.

Maguire, Bassett, Jr. 1963a. The passive dispersal of small aquatic organisms and their colonization of isolated bodies of water. Ecological Monographs 33:161-185.

Maguire, Bassett, Jr. 1963b. The exclusion of Colpoda (Ciliati) from superficially favorable habitats. Ecology 44:781-784.

Maguire, Bassett, Jr., Denton Belk, and Glenda Wells. 1968. Control of community structure by mosquito larvae. Ecology 49:207-210.

Maguire, Bassett, Jr. Aquatic communities in bromeliad leaf axils and the influence of radiation. In H. T. Odum (ed.), The tropical rain forest.

Maguire, Bassett, Jr. The limitation of trophic levels (in ms).

Paine, R. T. 1966. Food web complexity and species diversity. Amer. Naturalist. 100: 65-75.

Patrick, Ruth. 1971. Diatom communities. This symposium.

Slobodkin, L. B., F. E. Smith, and N. G. Hairston. 1967. Regulation in terrestrial ecosystems and the implied balance of nature. Amer. Naturalist. 101:109-124.

Snedecor, G. W. 1956. Statistical methods. 5th ed. Iowa State College Press, Ames. 534 p.

Tate, M. W. and R. C. Clelland. 1957. Non parametric and shortcut statistics. Interstate. Danville, Illinois.

Taub, Frieda B. 1969. A biological model of a freshwater community. Limnol. and Oceanog. 14:136-142.

Taub, Frieda B. 1971. A continuous gnotobiotic (species defined) ecosystem. This symposium.

Taub, Frieda B. and A. M. Dollar. 1968. The nutritional inadequacy of *Chlorella* and *Chlamydomonas* as food for *Daphnia*. Limnol. and Oceanog. 13:607-617.

Simberloff, D. S. and E. O. Wilson. 1969. Experimental zoogeography of islands. The colonization of empty islands. Ecology 50:267-278.

Silvey, J. K. G. 1971. Interactions between freshwater bacteria, algae, and fungi. This symposium.

Wilson, E. O. and D. S. Simberloff. 1969. Experimental zoogeography of islands. Defaunation and monitoring techniques. Ecology 50:278-296.

Ruth Patrick

Diatom Communities

Abstract

Results of studies of many diatom communities in streams have shown that these communities are typically composed of many species, most of which have relatively small populations, and the data fit the model of a truncated normal curve. The structure of the curve is very similar from season to season and from year to year. When one examines in detail the species composition, one finds that there are always some very rare species as well as those more common than the majority of species in a community. Some of the very rare may become more common under various seasonal conditions. However, some species never seem to be common. The role of these rare species in the community is not well understood, but it may be that their function is similar to that of rare genes in a gene pool—that is, to maintain the community over time during periods of great and unpredictable variation in environmental conditions.

A series of experiments was conducted to try to determine the effects of size of species pool capable of invading an area, frequency of reinvasion into an area, and the size of the area on the diversity of the community. The results of these studies indicate that the numbers of species capable of invading an area and the frequency of reinvasion are probably more important than actual size of area. Analyses of the data by various types of diversity indices have shown that these diatom communities as other stream communities have very low equability. One cause is probably the great variation in environmental factors that largely prevents species saturating the environment in which they live. Another contributing factor is the fact that nutrients rarely become limiting or waste materials autotoxic, for the nutrients are continually being replenished and the waste products continually removed. The effects of perturbation on these natural communities is discussed.

Introduction

Many studies have been made of the systematics and ecology of diatoms. Associations of diatoms have been used to determine water conditions since the latter part of the nineteenth century. Some of these studies were concerned with the patterns and influence of glaciation (Cleve, 1899, 1900; Cleve-Euler, 1940; Hyyppa, 1936; Hustedt, 1954; Hanna, 1933; Patrick, 1946; etc.). Other studies used diatoms to differentiate ocean currents (Cleve, 1892, 1899; Gran, 1897, 1900, 1902; Aikawa, 1936; Hendey, 1937). Diatoms have also been used to show changes in mineral content in lakes, rivers, and swamps (Hustedt, 1936; Kolbe, 1932; Cocke, et al., 1934; Patrick, 1936, 1968; Pennington, 1947; Ross, 1950). More recently diatom associations have been used to trace the nutrient levels and other changes in lakes in prehistoric times (Patrick, 1943, 1954; Pennington, 1947; Hutchinson, et al., 1956; Stockner and Benson, 1967; LaSalle 1966).

Some workers have tried to develop systems using indicator species or shifts in relative abundance of certain species to denote the degree of pollution of a body of water (Kolkwitz and Marsson, 1908; Nygaard, 1949; Foged, 1954; Cholnoky, 1958; Fjerdingstad, 1960; etc.).

Discussion

During the last twenty years the author has devoted a considerable amount of time to studying the structure of communities of aquatic organisms, particularly diatoms, in streams. In determining the structure of the community we are concerned with kinds of taxa, numbers of taxa, and relative sizes of their populations. The results of studies started in 1948 (Patrick, 1949, 1961) show that diatoms as well as communities of species of most major groups of aquatic life were represented by a fairly high number of species, many of which had moderate-to-small populations, although a few might be of common occurrence. Furthermore, the species composing the diatom communities changed considerably from season to season and from year to year, even though the structure of the community did not change very much.

Attempts were made in 1954 (Patrick, Hohn, and Wallace, 1954) to represent these relationships mathematically. It was found that a fairly large sample of a diatom community fitted the truncated normal curve as found by Preston for other groups of organisms (Preston, 1948). If one counted a sufficient number of specimens to always place the height of the mode in the same interval (i.e, the third interval to the right of the veil line), the shape of the curve as measured by height of the mode, σ^2, and intervals covered were remarkably stable from season to season and from year to year, if perturbation such as manmade pollution did not occur (Fig. 1, Table 1). It was found that the negative binomial also produced a fair fit, but the truncated normal curve was better for communities in natural continental streams.

The close similarity of the structure of natural diatom communities in streams enabled Patrick and Strawbridge (1963) to develop confidence intervals for σ^2 and the number of species in the mode for natural fresh soft water and brackish water streams. The normal bivariate distribution

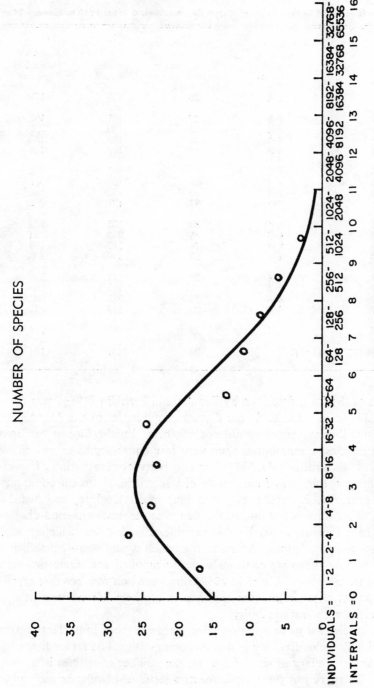

FIGURE 1. Graph of the diatom population for November 1951, from Ridley Creek, Chester County, Pennsylvania, a stream not adversely affected by pollution.

TABLE 1.—Structure of Savannah River Diatom Communities, October 1953 to January 1958.

Date	Specimen number in modal interval	Species in mode	Species observed	Species in theoretical universe
Oct. 1953	4–8	22	150	178
Jan. 1954	4–8	19	151	181
Apr. 1954	2–4	24	169	200
Jul. 1954	2–4	23	153	193
Oct. 1954	4–8	21	142	168
Jan. 1955	4–8	19	132	166
Apr. 1955	2–4	25	165	221
Jul. 1955	2–4	20	132	180
Oct. 1955	2–4	27	171	253
Jan. 1956	2–4	30	185	229
Apr. 1956	4–8	35	215	252
Jul. 1956	2–4	24	147	185
Oct. 1956	2–4	23	149	206
Jan. 1957	2–4	29	177	233
Apr. 1957	2–4	21	132	185
Jul. 1957	4–8	29	181	203
Oct. 1957	2–4	25	157	232
Jan. 1958	2–4	27	152	212
(Apr. 1954-1958 averages)		24	151	194

function of the form described by Bennett and Franklin (1954) was used.

Attempts were made to fit the diatom communities to the MacArthur Type I distribution, but they did not conform. Furtherfore, it has been shown that these communities have very low equability as measured by Lloyd and Ghelardi (1964). This low equability of distribution of specimens among species was found to be characteristic of stream communities of protozoa, invertebrates (including insects), algae, and usually for fish. This is probably due to the fact that the environmental characteristics of streams are unpredictably variable. Floods occur which greatly reduce species populations. Also a factor which would delay equability is the fact that nutrients are continually being supplied and waste products removed from an area. Lloyd in 1968 commented on the fact that reptile and amphibia populations that spend part of their life in streams had communities with low equability.

These studies led us to question what were the more important factors causing high species diversity in diatom communities. The factor first considered was the effect of size of area on the number of species in a community. To carry out these experiments a series of islands, or small glass squares supported on pedestals, was placed in a rack in a plastic box. The boxes were changed each day and the pedestals cleaned to assure that the only diatoms reaching the slides were from the water flowing over them. The structure of the diatom communities after various days of exposure indicates that the area size significantly affects the number of species composing the community. The numbers of species on the islands often increased to the first week, increased or decreased a little to the second week; and then subsequently remained about the same or slightly decreased, indicating that saturation as to numbers of species had been reached and some extinction was taking place that was not being replaced by invasion (Table 2). An examination of the sizes of populations of species composing the communities shows that populations represented by six or fewer specimens were those most subject to extinction.

A second series of experiments was performed in the stream from Roxborough spring, and a similar series was carried out in Ridley Creek. Previous studies have shown that, in the area of Roxborough spring stream being studied, about one hundred species of diatoms form the community at a given point in time, whereas in Ridley Creek the area studied supports about 250 species at a given point in time. The results of these studies clearly show that the same size area ($36mm^2$) supported a very different number of species, 14-29 in Roxborough spring and 160 in Ridley Creek. Thus it would appear that the potential species pool is more important than size of area in determining the number of species in a diatom community (Patrick, 1967).

The third series of experiments was to determine, if the area and species pool were the same, what would be the effect of varying the reinvasion rate. For these experiments a series of boxes (eight in number) was developed so that the boxes would support the same type of diatom communities (Patrick, 1968). It has been established that under similar flow rates (about 650 1/hr) 95% of the specimens were of the same species in all of the boxes. The rate of reinvasion of stream species was then varied (Fig. 2, Table 3), although the flow rate was maintained by recirculation. The results of experiments with an invasion rate of 1.5 1/hr show that the number of species composing the communities is less if repetition or frequency or reinvasion of the various species is reduced; that the communities consisted of fewer rare species, and that a greater proportion of the species had larger populations than under conditions of higher repetition of invasion.

Studies were then carried out in island streams of types similar to those we have studied in continental United States in the temperate zone. The main difference between the island streams and the streams in the United States was that the islands were tropical. If the theory that there

TABLE 2. (A)—Experiments in September-October, 1964.

	ROXBOROUGH SPRING				RIDLEY CREEK	
	625 mm² Slide		36 mm² Slide		9 mm² Slide	36 mm² Slide
	Number of Species				Number of Species	
	Box 1	Box 2	Box 3	Box 4	Box 6	Box 7
4 days	46	37	23	23	3	——
1 week	40	32	28	24	—	——
2 weeks	54	35	—	22	10	——
8 weeks	—	—	29	14	14	160

Table 2. (B)—Experiments in Roxborough Spring During Summer, 1964.

| | 1 Week 144 mm² Slide | | 2 Weeks 144 mm² Slide | | 1 Week 625 mm² Slide | | 2 Weeks 625 mm² Slide | |
	Box 1	Box 2	Box 3	Box 4	Box 5	Box 6	Box 7	Box 8
Number of species	32	28	23	22	47	44	29	28

TABLE 3.—Structure of Diatom Communities.

	Height of Mode	σ^2	Theoretical Number of Species	Observed Number of Species	Intervals Covered by Curve
Invasion Rate 550-600 1/hr. (Oct.-Nov., 1964) (Darby Creek)	22.4	6.2	140	123	9
Invasion Rate 1.5 1/hr. (Oct.-Nov., 1964) (Darby Creek)	15.3	12.0	133	97	15
Invasion Rate 550-600 1/hr. (Sept.-Oct., 1964) (Darby Creek)	22.5	6.9	148	129	9
Invasion Rate 1.5 1/hr. (Sept.-Oct., 1964) (Darby Creek)	13.9	12.6	124	100	12
St. Lucia—Canaries River	8.4	9.3	64	61	10
Island of Dominica stream (Layou River)	5.3	26.0	67	49	14
Island of Dominica stream (Check Hall River)	5.17	21.6	60	46	14
Maryland stream (Hunting Creek)	12.0	9.1	92	79	14

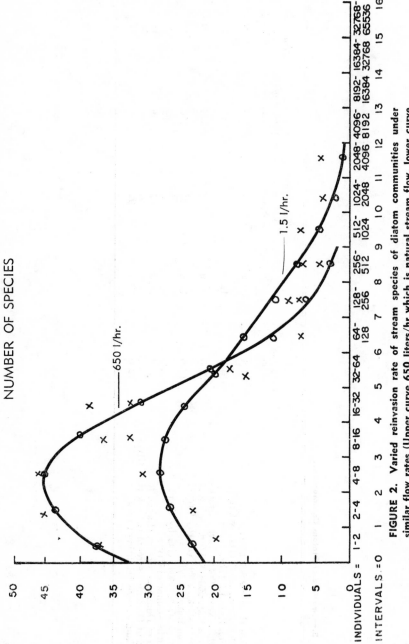

FIGURE 2. Varied reinvasion rate of stream species of diatom communities under similar flow rates (Upper curve 650 liters/hr which is natural stream flow, lower curve 1.5 liters/hr new water, recycled flow 650 liters/hr).

are larger species pools for diatoms in the tropics is correct, then one expects these islands to support more species than if they were temperate islands and the comparison with the temperate mainland would show less reduction due to isolation than if the islands were temperate islands.

However, the results of these studies do show the effect of isolation. The two streams in Dominica and the one in St. Lucia produced similar results. In all cases the numbers of species in the mode were small (Dominica, 5; St. Lucia, 8), and a greater percentage of the species composing the diatom communities had larger populations than one would expect in similar continental streams, usually resulting in σ^2 being larger (Table 3). The total number of observed species was also much lower, 46 to 61 as compared to 79 to 129 in continental streams.

Thus we see that size of area is important in maintaining a high number of species in a diatom community, but potential species pool capable of invading the area is more important. If reinvasion does not take place at a fairly high rate, diversity is reduced due to the extinction of the rarer species.

Thus the important factors in maintaining high species diversity in stream diatom communities seem to be the same as those for other groups of organisms (Hutchinson, 1952; MacArthur and Wilson, 1963). In natural continental streams a large species pool of diatoms is supplied from upstream or nearby stream areas. Since diatoms are usually dispersed by current, the turbulence and current in a river insures a high invasion rate as to number of species and reinvasion of the same species. Asexual reproduction and the rapidity of this reproduction (often once a day) insure a high birth rate and a short maturity time for reproduction. This method of reproduction also eliminates the necessity of finding a mate and reduces in part the necessity for cohesiveness in distribution. However, as pointed out by Nalewajko, et al. (1963) division or reproduction in algae is more likely to take place if a steady diffusion gradient of glycolate, which is liberated from algal cells, has been established around the cells. The presence of several cells close together would promote the establishment of a sufficient diffusion gradient.

Although the death rate is high due to predation and the precariousness of the environment, the factors mentioned above insure that normal continental communities can maintain high numbers of species and many of them with very small populations. If for any reason the invasion rate is significantly reduced the rapid reproductive rate may compensate, and similarly a high invasion rate may compensate for a low reproductive rate.

The importance of the maintenance of fugitive species or species with very small populations in a community should not be overlooked. They may be of great value in maintaining the community over time in an environment which is so variable. Indeed, their function in a community may be similar to that of rare genes in the gene pool of a species.

The effect of perturbation often caused by manmade pollution on diatom communities is to affect the colonization and population sizes of many species in the area and to often bring about a great change in the kinds of species that may live in an area.

The first effect of pollution is to affect rates of reproduction. As a result certain species, even if they are able to invade the area, are not able to reproduce and soon become extinct. Other species have their reproductive rate reduced, while the more tolerant species, because of less competition for nutrients, become more common. The shape of the truncated normal curve is altered as σ^2 becomes greater and Preston's "a" becomes less. The curve covers more intervals as some species become excessively dominant. This type of curve usually develops as eutrophication increases. A greater amount of organic pollution, high temperature, and some of the less toxic pollutants produce in addition to the above changes a reduction in numbers of species so that the height of the mode is less. There are relatively fewer species with small populations and some of the rare species with narrower tolerances are eliminated. Often species such as *Gomphonema parvulum* and *Nitzschia palea* which formerly had small populations greatly increase. Sometimes fairly large changes occur in the kinds of species that are common. Usually there are one or two species that are excessively common.

More severe toxic pollution may have one of the following effects. The invading species are not killed but cannot reproduce. In such cases the numbers of species may be high, σ^2 may be very small, and the number of intervals covered relatively few. The total biomass is very small. An example of this effect is caused when the pH is greatly reduced in a typical circumneutral stream (Patrick, et al., 1968).

Other types of toxic pollution greatly lower the height of the mode and the curve extends over only a few intervals. Sigma squared is large. This is the result of only a few relatively tolerant species surviving and having variable sizes of populations.

Thus it is apparent that the effect of perturbation may or may not destroy the invading species, but it usually alters the reproduction rate of the species in the area. As a result the more tolerant species, often the least desirable as a food source, increase. Perturbation may reduce or eliminate predator pressure by eliminating the predator or by reducing the desirability of the food source to the predator. Thus the population sizes of certain species increase.

Since the evenness of distribution of specimens among the species is greatly reduced, the Shannon-Weiner diversity index is lower. However, this index does not discriminate between certain types of changes. For example, the difference in the indices is relatively small between a community dominated by a few very common species, with only a few species with small to very small populations, and one with a few very common species, and many species with small to very small populations (Patrick, 1967; Sager and Hasler, 1969). The first type of community is characteristic of much more severe pollution than the latter.

In general, high diversity at the various trophic levels, as evidenced by numerous species with relatively small populations that are able to maintain themselves in associations over time, has been recognized by many as giving stability to the community, for it increases the flexibility of the community to respond to changing environmental conditions. High diversity is probably also important in predator-prey relationships, as laboratory experiments have shown that most organisms have better growth rates if fed on a mixed diet rather than a single food source. If a single species is used as a food source it is usually fortified with nutrients from other species. The importance of the maintenance of a diversified food source for the stability of a higher trophic level is shown by the work of Hairston, et al. (1968). It should also be noted that less entrophy is introduced into a system if the standing crop or population of a species is just large enough to insure continuance of the species over time, which means satisfying predator pressure, competition between species, and death due to density-independent factors. Thus natural communities typically have many species which are primary producers, herbivores, and omnivores, but most of them typically have moderate to small populations. Usually there are fewer carnivore species than species with other types of food preferences. If a large population of a species occurs it is soon cut down by predator pressure or severe changes in density-independent factors.

Perturbations due to man have upset these relationships so that large standing crops or nuisance growths result, which reduce the diversity of the species and change the energetics of the system. It is the meaning of these long-range changes, such as the decrease in the diversity of our ecosystems and the perturbations in the energy relationships of communities, that should be researched.

Literature Cited

Aikawa, M. 1936. On the diatom communities in the waters surrounding Japan. Rec. Oceanogr. Wks. in Japan, 8(1):1-160.

Bennett, C. and N. Franklin. 1954. Statistical analyses in chemistry and chemical industry. John Wiley & Sons, N.Y.

Cholnoky, B. J. 1958. Beitrag zu den Diatomeenassoziationen des Sumpfes Olifantsvlei südwestlich Johannesburg. Ber. Deutsch. Bot. Ges., 71(4): 177-187.

Cleve, P. T. 1892. Note sur les diatomees trouvées dans la poussiere glaciale de la coté orientale de Groënland. Le Diatomiste, 1:78.

—————. 1899. Postglaciala bildningarnas klassifikation pa grund af deras fossila diatomaceer. Sv. Geol. Und., ser. C, No. 180:59-61.

—————. 1900. The plankton of the North Sea, the English Channel, and the Skagerak, in 1898. Kongl. Svenska Vet.-Akad. Handl., 32(8):1-53.

—————-Euler, A. 1940. Das Letzinterglaziale Baltikum und de Diatomeenanalyse. Beih. Bot. Zentralbl., Abt. B, 60(3):287-334.

Cocke, E. C., I. F. Lewis and R. Patrick. 1934. A further study of Dismal Swamp peat. Amer. J. Bot., 21(7):374-395.

Fjerdingstad, E. 1960. Forurening af vandløb biologisk bedømt. Nord. Hyg. Tidskr., 41(7/8):149-196.

Foged, N. 1954. On the diatom flora of some Funen lakes. Folia Limn. Scand., 6:1-75.

Gran, H. H. 1897. Bemerkungen über das Plankton des Arktischen meeres. Ber. Deutsch. Bot. Ges., 15(2):132-136.

—————. 1900. Diatomacea from the ice-floes and the plankton of the Polar Sea. Norweg. North Pol. Exp. 1893-1896, Sci. Res., 3(11):1-74.

—————. 1902. Das Plankton des Norwegischen Nordmeeres, von biologischen und hydrografischen Gesichtspunkten behandel. Rep. Norweg. Fish. Mar. Invest., 2(2)5:1-222.

Hanna, G. D. 1933. Diatoms of the Florida peat deposits. Fla. State Geol. Surv., 23/24 Ann. Rep. (1930/1932):65-120.

Hairston, N. G., J. D. Allen, R. K. Colwell, J. D. Futuyma, J. Howell, M. D. Lubin, J. Mathias, J. H. Vandermeer. 1968. The relationship between species diversity and stability: an experimental appraisal with protozoa and bacteria. Ecology, 49(6):1091-1101.

Hendey, W. I. 1937. The plankton diatoms of the southern seas. Discovery Rep., 16:151-364.

Hustedt, F. 1936. Diatoms. Arch. Hydrobiol., 30(1):1-84.

—————. 1954. Die Diatomeenflora des Interglaziale von Oberohe in der Lünenburger Heide. Abh. Naturw. Ver. Bremen, 33(3):431-455.

Hutchinson, G. E. 1953. The concept of pattern in ecology. Proceedings, Academy of Natural Sciences of Philalelphia, 105:1-12.

—————, R. Patrick and E. S. Deevey. 1956. Sediments of Lake Patzcuaro, Michoacan, Mexico. Bull. Geol. Soc. Amer., 67:1491-1504.

Hyyppa, E. 1936. Über die spätquartäre Entwicklung Nordfinnlands mit Ergänzungen zur Kenntnis des spätglazialen Klimas. C. R. Soc. Geol. Finl., 9:401-465.

Kolbe, R. W. 1932. Grundlinien einer allgemeinen Ökologie der Diatomeen. Ergebn. d. Biol., 8:221-348.

Kolkwitz, R. and M. Marsson. 1908. Ökologie der pflanzlichen Saprobien. Ber. Deutsch. Bot. Ges., 26:505-519.

LaSalle, P. 1966. Lake quaternary vegetation and glacial history in the St. Lawrence lowlands, Canada. Leidse Geologische Mededelingen 38:91-128.

Lloyd, M. and R. J. Ghelardi. 1964. A table for calculating the equitability component of species diversity. J. Anim. Ecol., 33(2):217-225.

Lloyd, M., R. F. Inger and F. W. King. 1968. On the diversity of reptile and amphibian species in a Bornean rain forest. Amer. Nat., 102(928): 497-517.

MacArthur, R. H. and E. O. Wilson. 1963. An equilibrium theory of insular zoogeography. Evolution, 17(4):373-387.

Nalewajko, C., N. Chowdhuri, and G. E. Fogg. 1963. Excretion of glycollic acid and the growth of a planktonic *Chlorella*. *In* Microalgae and Photosynthetic Bacteria, 171-183.

Nygaard, G. 1949. Hydrobiologische Studien über dänische Teiche und Seen. II. The quotient hypothesis and some new or little known phytoplankton organisms. Kongl. Dansk. Vidensk. Selsk. Skr., 7(1):1-293.

Patrick, R. 1936. Some diatoms of Great Salt Lake. Bull. Torrey Bot. Club, 63(3): 157-166.

——————. 1938. A flora de quatro acudes da Parahyba. IV. Bacillariophyta. Ann. Acad. Brasil Sci., 10(2):89-103.

——————. 1943. The diatoms of Linsley Pond, Connecticut. Proc. Acad. Nat. Sci. Philadelphia, 95:53-110.

——————. 1946. Diatoms from Patzschke Bog, Texas. Not. Nat. Acad Nat. Sci. Philadelphia, No. 170:1-7.

——————. 1949. A proposed biological measure of stream conditions based on a survey of Conestoga basin, Lancaster County, Pennsylvania. Proc. Acad. Nat. Sci. Philadelphia, 101:277-341.

——————. 1954. The diatom flora of Bethany Bog. J. Protozool., 1:34-37.

——————. 1967. The effect of invasion rate, species pool, and size of area on the structure of the diatom community. Proc. Nat. Acad. Sci., 58(4):1335-1342.

——————. 1961. A study of the numbers and kinds of species found in rivers in eastern United States. Proc. Acad. Nat. Sci. Philadelphia, 113(10)215-258.

——————, M. H. Hohn and J. H. Wallace. 1954. A new method for determining the pattern of the diatom flora. Not. Nat. Acad. Nat. Sci. Philadelphia, No. 259:1-12.

Patrick, R. and D. Strawbridge. 1963. Methods of studying diatom populations. J. Wat. Poll. Contr. Fed., 25(2):151-161.

Patrick, R., N. A. Roberts, and B. Davis. 1968. The effect of changes in pH on the structure of diatom communities. Not. Nat. Acad. Nat. Sci. Philadelphia, No. 416:1-16.

Pennington, W. 1947. Studies on the postglacial history of British vegetation. Philos. Trans. Roy. Soc. London, 233:137-175.

Preston, F. W. 1948. The commonness, and rarity, of species. Ecol., 29:254-283.

Ross, R. 1950. Report on diatom flora from Hawks Tor, Cornwall. Philos. Trans. Roy. Soc. London, ser. B, 234(615):461-464.

Sager, P. E. and A. D. Hasler. 1969. Species diversity in lacustrine phytoplankton. I. The components of the index of diversity from Shannon's formula. Amer. Nat., 103(929):51-61.

Stockner, J. G. and W. W. Benson. 1967. The succession of diatom assemblages in the recent sediments of Lake Washington. Limn. Oceanogr., 12(3):513-532.

Ross E. McKinney

Microbial Relationships
in Biological Wastewater Treatment Systems

Abstract

With increased environmental pollution the importance of biological treatment systems cannot be over emphasized. Currently, there are four major treatment systems employing mixed microbial populations. There are three aerobic processes: trickling filters, activated sludge, and oxidation ponds; and there is one anaerobic process, anaerobic digestion. The aerobic processes are used where a high degree of treatment is required for dilute organic wastewaters. Anaerobic processes are used for treatment of concentrated organic wastes.

All of the biological treatment systems depend upon the interaction of various microorganisms under controlled environmental conditions. The bacteria are the major group of microorganisms responsible for waste stabilization with assistance from the fungi. Protozoa, rotifers, and crustaceans play a secondary role by consuming the excess bacteria, thereby producing a clarified effluent. Algae play only an incidental role in wastewater treatment.

While the biological sciences have contributed a limited amount of information on the microbiology of the wastewater treatment processes, they have not contributed significantly to the improvement of these biological treatment processes. Engineers are still designing, constructing, and operating biological wastewater treatment systems based on a trial-and-error methodology developed from past experience. Only with the activated sludge process has knowledge of microbiology and biochemistry resulted in significant improvement of the field designs and operations. The value of environmental pollution to the ultimate survival of mankind makes biological wastewater treatment systems an item of first priority for the biological sciences.

Introduction

With increased problems related to environmental pollution, there is greater concern than ever before for biological treatment systems that can purify wastewaters to a point where they can be discharged back into the environment without creating health hazards or nuisances. Unfortunately, the need for significant improvement in wastewater treatment has come so rapidly that the engineering profession has been unable to respond properly. The net result is simply that, while Congress and the people want pollution abated, it will not be abated in the near future.

Money and laws are not enough. Water pollution is simply the tragedy of neglect: neglect by the politician, neglect by the citizens, neglect by the engineers, neglect by the biologists.

Discussion

How can 100 million gallons of domestic sewage produced each and every day by a city of one million people be purified so that the water can be used again for other beneficial uses? What does a small midwestern town of 1,000 people do with its wastewaters after they have served their function?

People have discharged their waste materials into streams and lakes since the beginning of time. Yet, it has only been in recent time that people became concerned about the biological reactions created in the receiving waters as a result of waste discharges. The crowded island of Great Britain was one of the first to recognize the effects of failure to be concerned about stream pollution. Waring (1894) quoted from the 1858 Report on Sewage of Town Commissions in England, "That the increasing pollution of the rivers and streams of the country is an evil of national importance, which urgently demands the application of remedial measures." It appears that the situation has not changed significantly in the last 110 years. Stream pollution is still a national problem that demands a solution if mankind is to survive.

An ever increasing population and an exploding technology have pushed pollution ahead of wastewater treatment capabilities. Lots of progress has been made since 1858, but more is needed if this trend is to be reversed. Wastewater treatment systems have come a long way since the intermittent sand filter developed by Sir Edward Frankland in 1868. For the most part wastewater treatment systems have employed some form of biological treatment in order to obtain a high degree of stabilization of the organic pollutants. Microscopic examination of the microbial populations in the various wastewater treatment systems revealed a definite similarity with the microbial populations found in streams and lakes. The primary difference was one of numbers of microorganisms per unit volume of water. Under the controlled environments of the wastewater treatment units, the greater microbial populations produced a more rapid rate of reaction. In this manner it was possible to purify the wastewaters in a matter of hours in the treatment system as compared to days in the stream.

Unfortunately, the design and operation of wastewater treatment systems did not require a knowledge of the biological population dynamics. Treatment plants were built basically from previous experience. The intermittent sand filter gave way to the trickling filter in 1893 simply because the intermittent sand filter required too much land. In time the trickling filter was modified to accept greater and greater loads. The great advantage of the trickling filter lay in its simplicity of design and its ease of operation. So great was this advantage that the activated sludge process was largely ignored, even though it was discovered back in 1914 and produced a higher degree of treatment. Only the overwhelming pollutional load caused engineers to turn to activated sludge. Even then the design of the activated sludge system was based on past experience with little regard for the biochemical reactions within this treatment system. Only when engineers began to examine the microbiology of the biological treatment system was it evident that the currently accepted design criteria were not optimum. Significant modifications in activated sludge have been produced in recent years as a result of research on the microbiology and biochemistry of this process. There is no doubt that better operations and better designs for wastewater treatment systems will occur when engineers obtain a better understanding of the biological relationships within each treatment unit.

The Treatment Process

Wastewaters contain all of the materials which society can flush down the sewers. They contain domestic sewage as well as industrial wastes. Domestic sewage is largely feces, urine, food wastes, and wash waters, while industrial wastes include carbohydrates, proteins, fats, oils, alcohols, acids, phenolics, aromatics, salts, solids, and others. The value of the water carriage waste system is simply that the discarded wastes are carried away from their point of generation. Water flows steadily towards the oceans, moving on and on. Most people have little concern for their wastewaters once they have flowed past their property bounds. Only when the wastewaters generated by other people begin to create problems as they flow past the people downstream does anyone become concerned about wastewater treatment. There is no incentive for anyone to treat his own wastewaters, but there are definite incentives to see that his upstream neighbors treat their wastes properly.

Chemical examination of wasterwaters indicates that domestic sewage contains most of its organic matter as insoluble particles of varying sizes and most of its inorganic matter as soluble salts. Further examination of the organic fraction indicates that approximately 40 percent of the organic fraction is non-biodegradable. It is the biodegradable fraction that has created the major treatment problem in the past, but the inorganic salts are creating increasing problems and must also be reckoned with.

As sewage flows from its point of origin, bacteria begin to degrade the organic matter. As long as there is adequate dissolved oxygen, the bacterial metabolism is aerobic and reproduction is very rapid. Since oxygen is a very poorly soluble gas in water, it is not long before the oxygen is all removed from the wastewaters. Metabolism continues at a slow rate anaerobically. Anaerobic metabolism results in conversion of the solid organic particles into soluble organics. Mechanical turbulence of the sewage flow results in some disintegration. Thus, the longer the sewage flows the greater is the conversion from large insoluble organic solids to soluble organic solids and colloidal particles. It becomes immediately apparent that the treatment for any sewage will depend on the characteristics of the sewage when it reaches the treatment process. It is essential that engineers recognize this fact before they design the treatment plant. Failure to do so will result in a treatment plant that will not perform satisfactorily.

It should be recognized that wastewater treatment does not just consist of removing the pollutants from the water, but includes all processes involved until the pollutants are safely returned to the environment. Few people realize the complexity of total wastewater treatment systems but tend to think only about the original removal processes.

The first stage of sewage treatment is screening to remove the very large solids. These screenings can be ground and returned to the sewage or drained and buried. If the sewage contains much sand, it is passed through a grit chamber designed to remove the inorganic particles while allowing the organic particles to pass. The sand is usually washed free of organics and then used for fill material. The sewage then flows to a primary sedimentation tank where the heavy solids tend to settle out and the greases and oils tend to float. The heavy solids that settle must be removed and treated further. The greases and oils are skimmed from the surface and treated with the settled sludge or treated separately. In the past these concentrated solids have been treated by anaerobic biological digestion. The biodegradable organics were metabolised by bacteria with the production of methane and carbon dioxide, but the non-biodegradable organics and the inorganic solids required further processing prior to discharge back into the environment. Improved filtration techniques have permitted the direct dewatering of the primary solids and burial or incineration prior to burial of the ash. The effluent from primary treatment usually contains approximately 40% of the suspended solids and 65 to 70% of the BOD_5 contained in the raw sewage. Industrial wastes can change these values very significantly in either direction.

The primary effluent must be treated either in a trickling filter or in an activated sludge system. Trickling filters are nothing more than beds of rocks over which the wastes are discharged through a rotating distribution pipe. As the wastewaters flow over the stone surfaces, oxygen is removed from the atmosphere and rapid biological growth results. The biological population on the stone surfaces is sustained by the continuous addition of fresh wastes. The hydraulic washout causes some of the microbial population to be discharged from the rocks, preventing clogging of the filter. A secondary sedimentation tank follows the trickling filter in order to collect the excess microbial solids. These microbial solids must be handled in the same manner as the primary solids. In the activated sludge system the wastes are aerated to stimulate rapid aerobic growth. When an adequate microbial population has grown up, the microbes develop a tendency to form large floc, which will settle rapidly under quiescent conditions. As wastewaters are added to the aeration tank, microbes are displaced into a secondary sedimentation tank where separation occurs. The microbes are collected and returned to the aeration tank to maintain an adequate population for rapid metabolism. In time the system generates more microbes than can be maintained in the system and some must be wasted from the system. These excess microbial solids are handled along with the primary solids. The effluents from a trickling filter plant normally produce 85 to 90% removal of suspended solids and BOD_5, while activated sludge plants can produce from 90 to 99% removal of suspended solids and BOD_5.

In very small sewage treatment plants or in industrial wastes containing only soluble organics, the wastes are added directly to the aeration tank, eliminating the need for primary sedimentation tanks. Wastewaters are also added directly to oxidation ponds which permit sedimentation and stabilization in a single, earthen basin. Oxidation ponds have had widespread use in the midwestern part of the United States.

All of the biological wastewater treatment processes operate in the area of mixed microbial populations under varying environmental conditions. While there is a fixed pattern of microbial succession and while the biochemical changes within the wastewater system are related to this microbial pattern, little research has been done to develop sound concepts relating population dynamics to treatment plant results. By and large the biologists have looked to see which microorganisms were present; chemists have analyzed the treatment efficiencies; engineers have designed and constructed the treatment facilities; but few people have been concerned about the interrelationships of all of these factors. Yet, success depends upon both understanding and using the knowledge that each group has found.

Trickling Filters

Trickling filters in the United States tend towards being circular beds of rocks, 2 to 4 inches in diameter, stacked 6 feet high over a tile underdrain system that both collects the treated effluent and assists in distribution of air through the filter. The strongest organic wastes are deposited at the surface of the filter. As the waste travels through the filter, the microbial metabolism reduces the organic concentration so that there is less available food. Less food means less growth in the lower portion of the filter.

Initially, the filter is maintained aerobic by the thin film flow over the rocks. Bacteria respond readily to this aerobic environment and grow quite rapidly. Protozoa quickly respond to the good environment and feed on the bacteria. As the wastes are added to the filter, the bacteria continue to expand their population. Some of the bacteria are washed from the filter by the wasteflow; but some of the bacteria are retained in relatively quiescent zones where rocks make contact. Some of the bacteria form an attachment to the rock surface and soon the bacteria cover the entire stone. In the upper layers of the filter where the food concentration is relatively high, fungi tend to appear as do some filamentous bacteria. These filaments tend to give the bacterial mass a greater holding power. At the surface of the filter algae will usually appear where light can penetrate. Because of the thin layer of fluid the rich organic microbial layer is also attractive to worms and larvae. These higher forms of organisms are very important in reducing the bacterial mass in the filter.

As the bacteria grow out from the stone surface, oxygen does not penetrate below the upper layer. This causes the bacteria below the surface to shift to anaerobic metabolism. In time the bacteria at the stone surface die and the hydraulic force causes the bacterial growth to break off and wash out of the filter. The growth cycle then starts over again.

Betterfield and Wattie (1941) were the first to examine the effective bacteria in trickling filter slimes. They found that the predominant bacteria were zoogleal in nature. Wattie (1943) examined the cultural characteristics of these bacteria and classified them as belonging to the genus *Zoogloea*. They were satisfied that the zoogleal bacteria were responsible for the stabilization of the organic matter applied to the filter. On the other hand Cooke and Hirsch (1958) felt that fungi and algae were the basic microorganisms responsible for attachment to the stone surfaces. The most common fungi observed in trickling filters included *Fusarium aquaeductuum*, *Geotrichum candidum*, and *Pullularia pullulans*. There is no doubt that both bacteria and fungi contribute to the stabilization of the organic matter. The nature of the organic matter fed to the filter and the environmental conditions produced within the filter appear to be the factors determining the relative predomination of bacteria and fungi. It should be recognized that bacteria and fungi compete with each other for food. Generally, bacteria can compete more favorably for food than the fungi and will predominate in most trickling filters.

The protozoa in the trickling filter are found on the surface of the microbial films covering the stones. They are typical free-swimming ciliated and stalked ciliated protozoa that would be common for systems having high bacterial populations. Liebmann (1958) reported that *Colpidium colpoda* and *Glaucoma pyriformis* were common free-swimming ciliates in trickling filters; where *Vorticella* was a very common stalked ciliate. Later, Cooke (1959) summarized all the available data on the biology of trickling filters. His list of organisms observed in trickling filters was in excess of 200 species of bacteria, fungi, algae, protozoa, worms, and insects. This is the most complete list available to date. Yet little has been done to determine the interrelationships of all these organisms. The protozoa appear to be responsible for metabolising the bacteria which are not attached to the slime mass as well as those on the surface of the slime. The thin liquid layer on the surface of the microbial slime layer permits this liquid layer to be aerobic, creating a favorable environment for the protozoa. Since the protozoa are predators, they reduce the mass of microbial slime building up on the stones.

Worms and insect larvae appear to play a role in reducing the accumulation of slime in the trickling filter. Unfortunately, some of the higher animal forms have an adverse as well as a positive effect. The insect larvae eventually reach maturity and create nuisance conditions in the area around the trickling filter plant. Snail growths are so extensive in the Wichita, Kansas, trickling filters that they clog the sludge handling facilities following the filters.

Recent developments have used plastic trickling filter media instead of rock in order to keep the microbial film thickness to a minimum. In England alternating double filtration has been used to distribute the microbial growth more uniformly through the entire filter. The basic problem with the trickling filter is simply that the inert solids and dead microbial cells that accumulate in the filter must be removed if the filter is to continue to function properly. Generally, this is accomplished as a combination of hydraulic washout with death of the microbes at the stone surface. Since anaerobic conditions develop in the lower microbial depths, odorous materials will generally be released when the growth drops off the stones. For this reason trickling filters generally create odor nuisances in the immediate area surrounding the filters.

The physical and biological characteristics of a trickling filter prevent it from removing all of the biodegradable organics applied. Since the microbial growth is a function of food concentration, the growth decreases as the organics are removed. The net result is that there is little growth in the bottom of the filter if a high degree of treatment is obtained. Economics prevent building a filter tall enough to take out all of the organic matter.

Activated Sludge

The activated sludge process is fundamentally the most efficient biological wastewater treatment process. It operates as a dispersed microbial reactor under aerobic conditions to produce maximum stabilization of organic matter in the wastewaters being treated. The initial design, which employed long narrow aeration tanks, established a pattern from 1914 to the present. Engineers have shown almost no concern for the microbial relationships, being content to employ only hydraulic and other physical parameters to evaluate the treatment process. The net result was that while the system was capable of producing 90 to 95% BOD_5 reduction it was very sensitive and difficult to handle. This sensitivity retarded the use of activated sludge until recent times. Application of fundamental microbiological principles has resulted in some significant improvements in activated sludge systems and its application in wastewater treatment.

Buswell and Long (1923) were the first to report on the microbial changes that occurred during the formation of activated sludge. They found that when raw sewage was aerated, small flagellated and small ciliated protozoa were the first animal forms to appear. These soon gave way to larger ciliated protozoa. As the activated sludge began to form, the free-swimming ciliated protozoa gave way to stalked ciliated protozoa. Eventually rotifers appeared. From their observations they proposed a theory of activated sludge, but it had little impact on the engineers who designed activated sludge systems.

While Buswell and Long observed considerable bacterial activity, it was Butterfield (1935) who isolated the first floc-producing bacterium from activated sludge. It was felt that *Zoogloea ramigera* was the primary bacterium responsible for forming floc in activated sludge and was capable of stabilizing organic matter in the same manner as activated sludge. This observation established the bacterial basis of activated sludge.

Because protozoa were easy to observe and were obviously a part of the activated sludge system, efforts were made to determine the role that the protozoa played in the activated sludge process. Cramer (1931) observed that protozoa were an essential part of the activated sludge and that Vorticella could be used as an indicator of a well-operating activated sludge system. Heukelekian (1931) and Pillai and Subrahmanyan (1942, 1944) felt that protozoa were the primary agents of stabilization in activated sludge and that the bacteria played a minor role. Still this controversy had no effect on the design or operation of activated sludge plants.

It was recognized that if improvements in design were to be made, they would come only after engineers had a better understanding of the microbiology and biochemistry of the microorganisms making up the activated sludge. Research started at M.I.T. in 1950, culminated with the establishment by McKinney and Horwood (1952) that other bacteria than *Zoogloea ramigera* could form floc under the conditions in which activated sludge was formed. McKinney and Weichlein (1953) showed that most of the bacteria in activated sludge could form floc. There was no need for special slime-producing bacteria. Eventually McKinney (1956) set forth the energy theory of activated sludge formation which demonstrated the environmental conditions necessary for bacteria to form floc. This floc formation concept gave the treatment plant operators a practical mechanism for forming floc which worked every time. The relationships between protozoa and bacteria were studied by McKinney and Gram (1956). They showed that it was possible to produce a normal activated sludge comprised entirely of bacteria, although it produced a turbid effluent and lower BOD_5 reductions than normal activated sludge. Use of the flagellated protozoa, *Chilomonas paramecium* and *Euglena gracilis*, resulted in competition between the bacteria and the protozoa for food. The protozoa soon lost out and the systems having the protozoa seed soon reacted as the bacteria control. On the other hand the ciliated protozoa, *Tetrahymena gelii* and *Glaucoma scintillans*, used the bacteria for their substrate and survived. These two protozoa units quickly formed normal activated sludge with a clarified effluent. It was apparent that the protozoa were metabolising the dispersed bacteria. This helped establish definitely that the bacteria are the primary organisms of stabilization and that the protozoa have a secondary role in clarification. Both functions are extremely important if a high degree of treatment is to be obtained. It was later observed that protozoa were excellent indicators of aerobic conditions as well as indicators of toxic conditions. Thus, the treatment plant operator obtained some very simple operational guides from fundamental microbiological studies.

Further studies on the biochemistry of activated sludge at M.I.T. revealed that conventional activated sludge was not properly designed from a biological standpoint. It was found that the bacteria were in a feed-starve cycle which made the activated sludge system very sensitive. It was never in equilibrium with the organic load imposed on the microorganisms. The net result was operational problems when the load varied significantly from the average. In an effort to keep the microorganisms in equilibrium with the incoming organic load and to level out the organic fluctuations, McKinney (1956) recommended the use of complete mixing in the aeration tank. Complete mixing activated sludge was new and had seen little use. To demonstrate the value of completely mixing activated sludge, a number of treatment plants were constructed which treated both domestic sewage and a wide variety of industrial wastes. The results from these treatment plants and many others have demonstrated that complete mixing activated sludge is superior to conventional activated sludge. This was the first significant improvement in biological waste treatment systems that had its origin in understanding the ecological changes occurring in these systems and applying that knowledge to optimize treatment system design. The net result is a higher degree of treatment than possible before at less cost.

In recent years considerable research has been carried out with regard to the microbiology of activated sludge as well as on the floc-forming characteristics. While these studies have added additional information as to the types of bacteria in activated sludge and some data on the role that cations play in flocculation, none of them have contributed significant information to help in the design and operation of activated sludge systems.

Anaerobic Digestion

The concentrated organic matter in the primary sludge as well as the waste microbial solids from the trickling filters and the activated sludge process must undergo stabilization before returning to the environment. It was found that these concentrated organics could best be stabilized by anaerobic biological systems. Investigation of the anaerobic systems indicated that predominant microorganisms were bacteria. The operating characteristics of the anaerobic digestion system indicated that there were two major groups of bacteria, acid-forming bacteria and methane-forming bacteria. The first group of bacteria was believed to break down the complex organic solids to soluble organic acids. The methane bacteria metabolized the organic acids to methane and carbon dioxide. Very little work has been done on the microbiology of anaerobic systems, since it is so difficult to work in an anaerobic environment. Most of the research has been done on mixed microbial systems.

Maki (1954) found that primary sewage sludge contained approximately 34% cellulose. He isolated and studied 10 strains of bacteria which could metabolize cellulose under anaerobic conditions. These cellulose bacteria were not identified. Cooke (1965) found that fungi could survive for long periods in anaerobic digesters but he was not able to ascertain if the fungi participated significantly in the stabilization of the organic matter. In view of the low populations reported, it is doubtful if the fungi contributed significantly to the stabilization of the organic matter. Toerien (1967) isolated a number of aerobic and facultative bacteria from an anaerobic digester fed sewage sludge. These bacteria belonged to the genera *Bacillus, Micrococcus,* and *Pseudomonas.* Mah and Sussman (1967) found that the bacteria from an anaerobic digester produced 10 times more organisms when grown anaerobically than aerobically. This was the first report that indicated the acid formers were strict anaerobes rather than facultative.

The methane bacteria have received some attention. Barker (1956) reported on nine species of methane-producing bacteria belonging to the genera *Methanobacterium, Methanococcus,* and *Methanosarcina.* To date there has been little effort to study the methane bacteria under controlled conditions such as are required in an anaerobic digester. There is much work to be done in this area if the anaerobic process is to be improved. Failure to produce significant improvement in the anaerobic process will result in its displacement by chemical and physical treatment processes.

Occasionally protozoa have been observed in anaerobic digesters; but the numbers of protozoa are so small that they do not contribute to the process. It appears that the low numbers of protozoa are related to the low bacterial populations in these systems. It is hoped that there will be increased interest in developing a basic understanding of the microbiology of anaerobic digestion so that it can be more effectively applied in waste treatment systems. The real advantage of the anaerobic system is the production of a usable end product, methane. A secondary advantage is the low microbial solids production, hence less residual solids.

Oxidation Ponds

One of the most controversial of the waste treatment processes in use today is the oxidation pond. It is the simplest and most economical form of wastewater treatment. Yet it is one of the most misunderstood processes.

The basic concept of oxidation ponds was one of symbiosis between the bacteria and the algae. Oswald and Gotaas (1955) pictured the reaction as one in which the bacteria aerobically stabilized the organic matter and produced end products which could be used by algae to produce the oxygen needed by the bacteria for the stabilization process. There is no doubt that this was generally correct, but it fell short of the entire picture.

One of the first scientific studies on oxidation ponds was by Wenn-strom (1955) at Lund, Sweden. This treatment plant employed primary sedimentation of its raw sewage and four oxidation ponds in series. This study included both chemical and biological analyses. The major part of the biological analysis lay in identification of the predominant protozoa in each pond. He found that the free-swimming ciliated protozoa *Glau-coma scintillans* eventually gave way to *Vorticella microstoma* much in the same way as in activated sludge. Rotifers and *Daphnia* occurred in Pond IV in the summer months and consumed most of the excess algae.

The primary push for oxidation ponds came from the Kansas City office of the U.S.P.H.S. Neel and Hopkins (1956) set the stage for extensive use of oxidation ponds with a detailed study at Kearney, Nebraska. Extensive lists of phytoplankton and zooplankton were made. The predominant phytoplankton were *Chlamydomonas, Micractinium,* and *Eu-glena.* Ciliated protozoa and rotifers predominated for the zooplankton. The algae were found to be the prime source of the oxygen in the pond and bicarbonates were the source of carbon dioxide for the algae. It was found that the effluent contained very high concentrations of plankton during the warm months.

Bartsch and Allum (1957) looked beyond the simple symbiotic relationship and found that the oxygen demand in a sewage-fed oxidation pond due to the sewage stabilization was only 6% of the total respiratory oxygen demand. It seems that the algae themselves were the primary organisms demanding the oxygen produced by photosynthesis. This study raised lots of questions as to what was really happening in oxidation ponds, but did not follow up on them. The net result was simply that oxidation ponds continued to be constructed as before in ever increasing numbers. It is apparent that the extensive algal growth resulted in the production of more organic matter than was contained in the raw sewage. Discharge of the plankton in the effluent did not solve the pollution problem, but merely changed its form.

Summary

The four basic forms of biological treatment follow the same patterns. The organic matter in the wastes is stabilized by the bacteria best able to survive in the environment imposed by the specific treatment process, whether aerobic or anaerobic. The nature of the organics, their concentration, nutrient elements, trace elements, pH, temperature, time, and ultimate hydrogen acceptor are all factors that determine predomination. The various species of bacteria compete with each other for food. Fungi also compete with bacteria for food, but can predominate only when the environment contains only a slight concentration of dissolved oxygen, a low pH, or a low nitrogen concentration. Fungi are undesirable in activated sludge systems because the filaments prevent the activated sludge from settling properly in the secondary sedimentation tank. This is not a problem in trickling filters, but the degree of purification in trickling filters is not as great as in activated sludge. Bacteria can essentially remove all of the biodegradable organics contained in the raw wastes. The most unique property of the bacteria is their ability to flocculate once the organics in the wastes have been stabilized. Flocculation greatly assists in the removal of both the non-biodegradable solids and the micro-organisms.

Protozoa, rotifers and crustaceans are responsible for metabolism of excess dispersed bacteria, thereby producing a clarified effluent. The specific predomination is a function of the microbial plants and the environment. Basically, any efficient biological wastewater treatment system depends upon maintenance of the proper balance between microbial plants and microbial animals.

Unfortunately, the production of a highly stable, clarified effluent has produced a situation for massive plant growth on the inorganic nutrients remaining. Massive growths of water hyacinths in Florida have clogged inland water canals. Massive growths of algae in lakes have resulted in accelerated destruction of the useful value of those lakes. This poses the need for a new form of treatment, a tertiary stage. Currently, tertiary treatment is chemical and physical, but there is indication that it could be biological if someone understood how to apply the basic fundamentals of algae metabolism to the design of the proper treatment system. Laboratory results have demonstrated that it is possible to produce an activated algae much like activated sludge, but much work needs to be done to produce a practical system. This is the challenge of the future.

Conclusions

To date biological research has contributed bits and pieces of data related to various biological wastewater treatment systems, but, with one exception, has failed to make a significant contribution to the improvement in design or operation of biological wastewater treatment systems. A significant improvement was made in the design and operation of activated sludge systems from fundamental knowledge of the microbiology and the biochemistry of the basic process. With environmental pollution increasing at an accelerating rate, the need for improved biological wastewater treatment systems is overwhelming. Unless massive efforts are made to develop new systems in the immediate future, there is little hope that the challenge of environmental pollution will be met.

Literature Cited

Barker, H. A. 1965. Bacterial fermentations. John Wiley & Sons, New York.

Bartsch, A. F. and M. O. Allum. 1957. Biological factors in treatment of raw sewage in artificial ponds. Limnol. and Oceanog. 2, 77-84.

Buswell, A. M. and H. L. Long. 1923. Microbiology and theory of activated sludge. Illinois Water Survey Bulletin No. 18, 82-92.

Butterfield, C. T. 1935. Studies of sewage purification II. A zooglea-forming bacterium isolated from activated sludge. Public Health Reports, 50, 671-684.

Butterfield, C. T. and E. Wattie. 1941. Studies of sewage purification XV. Effective bacteria in purification by trickling filters. Public Health Reports, 56, 2445-2464.

Cooke, W. B. and A. Hirsch. 1958. Continuous sampling of trickling filter populations II. Populations. Sewage and Ind. Wastes, 30, 138-156.

Cooke, W. B. Trickling filter ecology. Ecology, 40, 273-291.

—————————. 1965. Fungi in sludge digesters. Proceedings of the 20th industrial waste conference. Purdue University, 6-17.

Cramer, R. 1931. The role of protozoa in activated sludge. Ind. and engr. chem., 23, 309.

Heukelekian, H. 1931. Partial and complete sterilization of activated sludge and the effect on purification. Sewage Works Journal, 3, 369-373.

Heukelekian, H. and M. Burbaxani. 1949. Effect of certain physical and chemical agents on the bacteria and protozoa of activated sludge. Sewage Works Journal, 21, 811-17.

Liebmann, H. 1958. Uber die Existenzbedingungen der Organismen im Tropfkorper und im Belebungsbecken. *Tropfkorper and Belebungsbecken*. R. Oldenbourg, Munich, 11-20.

McKinney, R. E. and M. P. Horwood. 1952. Fundamental approach to the activated sludge process I. Floc producing bacteria. Sewage and Ind. Wastes, 24, 117-123.

McKinney, R. E. and R. G. Weichlein. 1953. Isolation of floc producing bacteria from activated sludge. Applied Microbiology, 1, 259-261.

McKinney, R. E. 1956a. Biological flocculation. Biological treatment of sewage and industrial wastes, Vol. 1, Aerobic oxidation. Reinhold Publishing Co., 88-100.

—————————. 1956b. The use of biological waste treatment systems for the stabilization of industrial wastes. Proceedings of the 11th industrial waste conference. Purdue University, 465-477.

McKinney, R. E. and A. Gram. 1956. Protozoa and activated sludge. Sewage and Industrial Wastes, 28, 1219-1231.

Mah, R. A. and C. Sussman. 1967. Microbiology of anaerobic sludge fermentation I. Enumeration of the nonmethanogenic anaerobic bacteria. Applied Microbiology, 16, 358-361.

Maki, L. R. 1954. Experiments on the microbiology of cellulose decomposition in a municipal sewage plant. Antonie van Leeuwenhoek, 20, 185-200.

Neel, J. K. and G. J. Hopkins. 1956. Experimental lagooning of raw sewage. Sewage and Industrial Wastes, 28, 1326-1356.

Oswald, W. J. and H. B. Gotaas. 1955. Photosynthesis in sewage treatment. Proceedings American Society of Civil Engineers, 81, Separate 686.

Pillai, S. C. 1941. The function of protozoa in activated sludge process. Current Sci. (India), 10, 84-85.

Pillai, S. C. and V. Subrahmanian. 1942. Role of protozoa in activated sludge. Nature, 150, 525.

Pillai, S. C. and V. Subrahmanian. 1944. Role of protozoa in the aerobic purification of sewage. Nature, 154, 179-180.

Toerien, D. F. 1967. Direct-isolation studies on the aerobic and facultative anaerobic bacterial flora of anaerobic digesters receiving raw sewage sludge. Water Research, 1, 55-59.

Waring, G. E. 1894. Modern methods of sewage disposal. D. Van Nostrand Co., New York.

Wattie, E. 1943. Cultural characteristics of zooglea-forming bacteria isolated from activated sludge and trickling filters. Public Health Reports, 57, 1519-1534.

Wennstrom, M. 1955. Oxidation ponds in Sweden. Lunds Universities Arsskrift. N.F. Avd. 2 Bd 51 Nr 7.

John E. Hobbie

Heterotrophic Bacteria
in Aquatic Ecosystems;
Some Results
of Studies with Organic Radioisotopes

Abstract

The rates at which heterotrophic bacteria transform organic material are virtually unknown in aquatic ecosystems. One experimental approach is to add very small quantities of radioactive organic compounds and measure the uptake into the bacteria after short incubations. The kinetics of the uptake can be studied by adding different concentrations of substrate and analyzing by the Michaelis-Menten formula. Maximum velocities of uptake are highest after spring and late summer algal blooms and may drop to a tenth of those values during the winter. The concentrations of the substrates, such as glucose, acetate, or amino acids, remain at low levels throughout the year as the rate of supply to the dissolved organic pool appears to be balanced by the rate of removal by the bacteria. In polluted systems, the bacteria may remove all of the substrate within ½ hour; in oligotrophic systems it may take over 1000 hours. When respiration is calculated and the actual concentration of substrate known, the flux of carbon through these bacteria can be estimated. In one estuarine system, the flux of amino acids alone accounted for close to 10% of the primary productivity. In eutrophic systems the bacteria can exist upon the dissolved organic carbon and are always actively taking up substrate. In oigotrophic systems, the transport systems have to be induced before uptake of the substrate occurs.

Introduction

Heterotrophic bacteria are probably the most ecologically important type of bacteria occurring in most lakes, yet the rates at which they transform organic material are virtually unknown. Plate count estimations of their populations, which have been made for many years, tell little about the heterotrophic activity within the water column. One of the methods that has given some information on the dynamics of the ecosystem is a measure of the uptake of organic radioisotopes. This paper presents some results and insights from the use of this method and reviews some important roles of the heterotrophic bacteria.

One little understood role of heterotrophic bacteria in lakes is as concentrators of nutrients, particularly of nitrogens and phosphorus. This

effect is well understood in soil fertility studies, where addition of organic
material poor in nitrogen or phosphorus actually can reduce the productivity of a field. The bacteria decomposing this organic material need
a C:N:P ratio of about 125:5:1 in order to grow—the necessary N or P
had to come from the soil (Alexander, 1961). This effect of the bacteria
on nutrients has not been widely appreciated in aquatic biology, but an
example of photosynthetic limitation in estuaries comes from the work
of Thayer (1969). In these experiments (Fig. 1), the uptake of carbon-
14 (as bicarbonate) in the light was measured over time for subsamples
with no added nutrient (control), with all nutrients added, and with all
nutrients plus glucose or chopped *Spartina* (cord grass). When glucose
was added, the bacteria multiplied and removed so much of the N and P
from the water that the photosynthetic algae did not show an increased
uptake of carbon-14 relative to the controls. The *Spartina* had the same
effect on the algal photosynthesis, but it should be noted that the effects

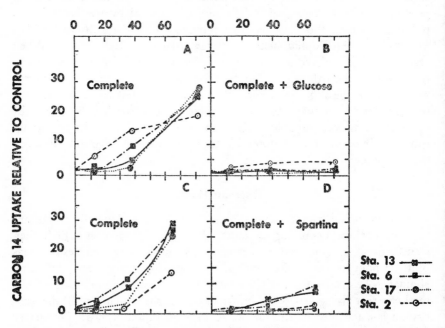

FIGURE 1. Carbon-14 uptake of phytoplankton relative to uptake of controls during
summer, 1968, in the shallow estuaries near Beaufort, North Carolina. The controls
received no enrichment while the complete samples received all the nutrients. Glucose was added to one series (125 ug-at glucose C/liter) and chopped *Spartina* (25
mg/liter) to the other. Modified from Thayer (1969).

of enclosing the samples in bottles and of long incubations (up to 80 hr) make it difficult to apply these results directly to natural estuaries.

Another way that heterotrophic bacteria can be extremely important in aquatic ecosystems is through the formation of particulate material. This, of course, is occurring all the time, but where a large amount of organic matter enters the system from outside (allochthonous) the formation of particulate carbon by bacteria can exceed that by algae. The example given in Fig. 2, from Kuznetsov (1968), is of a reservoir. Nauwerck (1963) calculated that the algal production was not enough to support the zooplankton in Lake Erken and postulated that bacterial breakdown of macrophytes and blue-greens was very important in providing the extra food necessary. In 1964-65, several years after Nauwerck's investigation, the dissolved organic carbon was measured in the same lake (Fig. 3). The annual cycle will be discussed later, but a net decrease of some 3 mg C/liter was found over the year. This amount is approximately 60 gm C/m²/year or about one-quarter of the annual algal productivity. It is, of course, a very conservative estimate of bacteria heterotrophy.

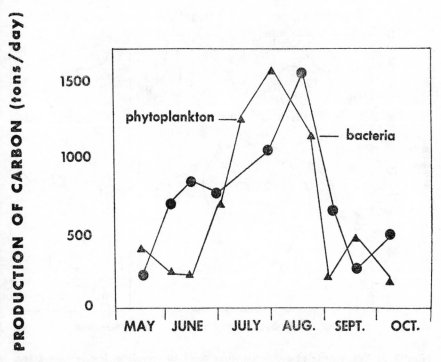

FIGURE 2. Daily production of bacteria and algae in Rybinsk Reservoir, 1964 (modified from Kuznetsov, 1968).

Actually, there may be a very close relationship between the excretion of dissolved organic matter by aquatic plants and the uptake of this material by bacteria. If the bacteria are quite active, the dissolved organic matter would be taken up as fast as it was excreted and there would be little total build-up. Estimations of total dissolved organic matter passing through the bacteria would therefore be seriously underestimated (for example, it is likely that the data in Fig. 3 should have been collected on a daily or hourly basis instead of once a month). Evidence is building up that both phytoplankton algae (Watt and Fogg, 1966) and rooted aquatics (Wetzel, 1969) excrete a sizeable percentage of the photosynthetically fixed carbon as dissolved organic matter. One excreted compound has been identified as glycollic acid and Wright (in press) found active bacterial uptake of this compound only in the upper layers of a pond (Table 1). Evidently the bacteria are using at least some of this material.

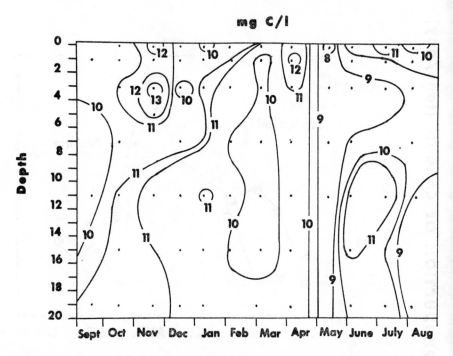

mg C/l

FIGURE 3. Yearly cycle of concentration of dissolved organic carbon in Lake Erken, Sweden, during 1964-65. Determinations made with infra-red analysis. Unpublished data of J. E. Hobbie and R. T. Wright.

TABLE 1.—Rates of Photosynthesis, Excretion of Dissolved Organic Matter, and Bacterial Uptake of Glycollate (heterotrophic potential) in Gravel Pond, Massachusetts. The algal photosynthesis was measured with carbonate-C^{14}, the excreted carbon from the amount of C^{14} in the dissolved organic material, and the bacterial uptake by measuring the maximum velocity of glycollate-C^{14}. Modified from Wright (in press).

Depth	Photosynthesis	Excreted material	Glycollate uptake
(m)	(ug C/liter/day)	(ug C/liter/day)	(ug C/liter/day)
0	134	7.8	6.7
1	120	11	5.3
3	82.3	2.5	5.8
5	24.8	2.4	1.5
7	6.7	3.2	0.7

The technique described here depends upon the fact that the bacteria take up organic compounds by a series of specific transport systems and the kinetics of this uptake can be described by equations worked out for enzymes. Thus, when the uptake of glucose-^{14}C by planktonic bacteria in a water sample is tested at a series of different concentrations of added substrate, A (μg/liter), a maximum velocity of uptake, V (μg/liter/hr), is soon reached. Ideally, the measurement of interest is v, the actual uptake velocity at the natural substrate concentration (S). However, in natural waters the glucose and other compounds are each found at a concentration of only a few μg/liter, too low to measure in a routine fashion. Because v depends upon S, it cannot be easily measured. Luckily, a measure of V can be made as well as two other parameters of ecological interest.

The equation used is a modified Lineweaver-Burk form of the Michaelis-Menten equation whose derivation in this context is described by Wright and Hobbie (1966).

$$\frac{t}{F} = \frac{K+S}{V} + \frac{A}{V}$$

The fraction of the isotope taken up by the bacteria (f) divided into the time of incubation (t) is a turnover time for the total substrate (S + A). Results of t/f for a series of experiments at different levels of A can be plotted against A giving a straight line. Extrapolation of the resulting straight line to A=O, gives the turnover time (T) when only S is present (the natural turnover time (hr) of the particular substrate in nature). The reciprocal of the slope equals V and a value for (K + S) can also be calculated from the ordinate intercept (K is a constant representing substrate concentration at one-half V).

The results of an experiment where four concentrations of aspartic acid-C^{14} were added to pond water and incubated for 2 hr at 10°C are given in Fig. 4. The radioactivity retained by the bacteria after filtration onto a Millipore filter (0.45 μ pore size) is called net uptake. From this plot of t/f against A, a V of 0.028 μg aspartic acid/liter/hr, a T (turnover time) of 131 hr, and a (K + S) of 92 μg/liter were calculated.

The V gives a measure of potential heterotrophy; that is, the velocity of uptake when the transport system is saturated. If the bacteria reproduce or induce more uptake sites, the V will increase proportionately. Thus, V is both an indication of what the bacteria are doing and an approximation of their level of activity. The actual level of uptake (v) in the water is much less, perhaps between 10 and 50% of the V.

The turnover time (T) of the natural substrate (S) is the number of hours required for the bacteria to remove all of the S from the water. Defined as S/v, it appears to vary by orders of magnitude mainly because of changes in v and is, therefore, another indicator of heterotrophic activity. More important, the T suggests that there is a pool of small-chain organic molecules that are being removed from the dissolved organic matter pool quite rapidly. Although it is difficult to prove, it is likely that most of the dissolved organic carbon is in the form of long-chain compounds that are only slowly affected by the bacteria.

Additional evidence for low concentrations of S comes from the (K + S) measurement. Although in the experiments described so far it is not possible to separate these constants, the sum of the two gives a maximum value for S. As noted above, the (K + S) does not change appreciably over the year.

The data for radioisotope uptake presented here do not take into consideration any carbon-14 lost by respiration. A relatively simple method of collecting this $^{14}CO_2$ has been described by Hobbie and Crawford (1969) and uses a filter paper wick moistened with a basic amine. After the $^{14}CO_2$ is taken up on the wick, the contained activity is counted by liquid scintillation. The amount of respiration varies with the compound and population, but it can reach 60% of the total activity taken up by the bacteria. One example is the aspartic acid uptake experiment (Fig. 4) where the corrected curve (net + respiration) has a V some 60% higher than the net curve.

Results and Discussion

Chlorophyll *a* (Fig. 5) and particulate carbon data (Fig. 6) from Lake Erken reflect the algal productivity peaks that occurred in May immediately after the ice left the lake and in late August and early September.

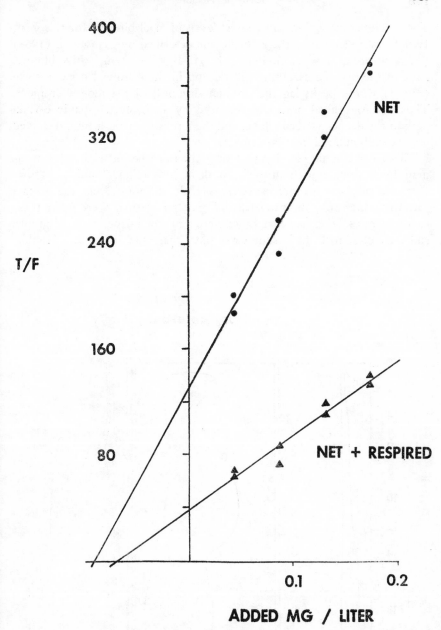

FIGURE 4. The uptake of aspartic acid by bacteria. In this plot (modified Michaelis-Menten) the ordinate units are hours. The upper line is calculated from isotope remaining in the bacteria (net) while the lower line is the net plus the amount of isotope respired. Sample collected from the Dairy Pond, Raleigh, North Carolina on 22 March, 1968.

The species composition and general levels of algal biomass for this year, 1964-65, were similar to the annual cycle described by Nauwerck (1963) (A. Nauwerck, personal communication). In spite of impressive blooms, most of the organic carbon is still tied up in the dissolved organic form (Fig. 3) whose peaks lag the productivity peaks by a month or more. Thus, the May algal peak was followed by a dissolved organic carbon peak in the deeper water in June, and the late August peak by a dissolved organic carbon buildup in November.

The heterotrophic activity (V) over the same period reached a maximum in September with an earlier peak in June (Figs. 7 and 8). However, the patterns for glucose were slightly different from the acetate patterns, which may indicate that different populations were using these two substrates. The turnover times were a mirror image of the V graphs and were close to 10 hr in summer to more than 1000 in winter (Hobbie,

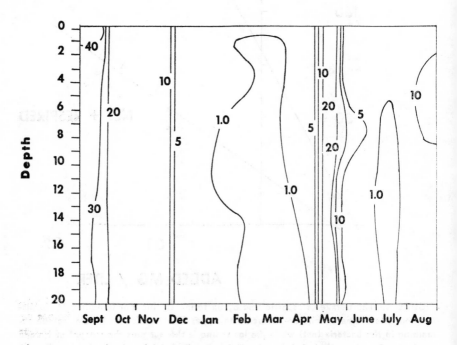

FIGURE 5. The yearly cycle (1964-65) of chlorophyll *a* in Lake Erken, Sweden. Unpublished data of R. T. Wright and J. E. Hobbie.

1967). The (K + S) values were always close to 5 μg/liter for glucose and close to 10 μg/liter for acetate. With these two substrates, it is likely that the S is about 50% of the (K + S). If this is so, then the flux of dissolved glucose and acetate into the bacteria is 8 gm C/m^2/yr. These substrates are only two of the many sugars, amino acids, and fatty acids that could be taken up by bacteria—this suggests that the 60 gm C/m^2/yr derived from changes in the dissolved organic carbon is a very conservative estimate of the total flux through the bacteria.

In more recent studies, the actual concentration of substrate (S) was measured separately along with the respired $^{14}CO_2$ to give flux of amino acids in the York River estuary (Hobbie, Crawford, and Webb, 1968). In the Pamlico River, also an estuary, the total flux of 11 amino acids was 88 mg C/m^2/day (Table 2). This table also illustrates that the K values for amino acids are always much higher than S (this was not the case with glucose or acetate) and that much of the amino acid taken up is immediately respired.

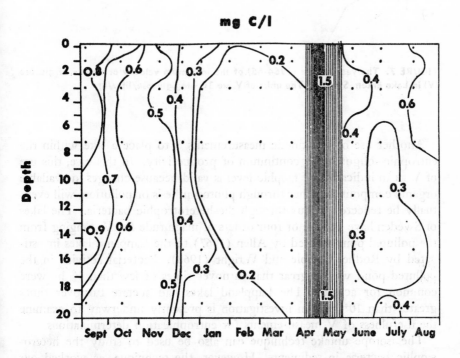

mg C/l

FIGURE 6. The yearly cycle (1964-65) of particulate organic material in Lake Erken, Sweden. The particles were collected on glass fiber filters and analyzed with a wet oxidation and spectrophotometric measurement.

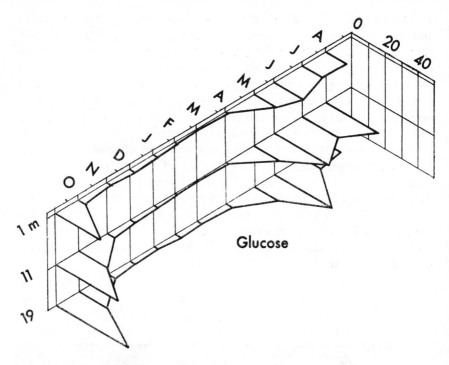

FIGURE 7. The yearly cycle (1964-65) of the maximum velocity of uptake of glucose (V) in Lake Erken, Sweden. The units of V are 10^{-5} mg glucose/liter/hr.

Another use of the kinetic measurements is to place a lake within the eutrophic—oligotrophic continuum of productivity. In this case, this use of V as an indication of trophic level is valid because changes in available organic compounds, either through photosynthesis or pollution, will eventually be reflected in flux through the heterotrophic bacteria. The lakes of Sweden have a range of four orders of magnitude for V, ranging from the polluted pond studied by Allen (1967) to the Lappland lakes investigated by Rodhe, Hobbie and Wright (1968). Bacterial activity in the polluted pond was so great that turnover times of less than ½ hr were common for acetate. The Lappland lakes had acetate turnover times greater than 10^4 hr. An investigation is presently underway to examine the usefulness of V measurements for eutrophication determinations.

The isotope uptake technique can also be used to study the heterotrophic bacteria in sediments. However, the technique, as worked out by Lindsay W. Wood (unpublished) must be modified by diluting the sample with artificial sea water (100 times) because of the intense bac-

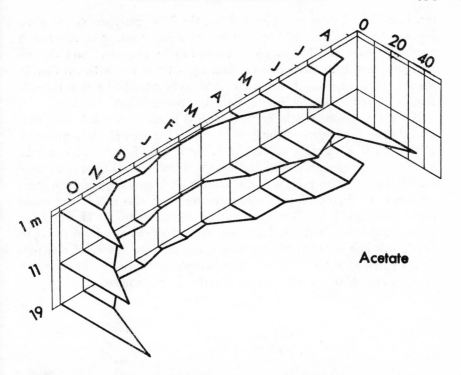

Acetate

FIGURE 8. The yearly cycle (1964-65) of the maximum velocity of uptake of acetate (V) in Lake Erken, Sweden. The units of V are 10^{-5} mg acetate/liter/hr.

TABLE 2.—Uptake Parameters and Concentrations of Dissolved Free Amino Acids in a Pamlico River (North Carolina) Sample Taken on March 25, 1969. The substrate concentration (S) was determined by K. L. Webb and the bacterial uptake by C. C. Crawford. All values of V, v, and T were corrected for respiration and the percentage respiration is (radiosotope respired) (total radioisotope entering cell)$^{-1}$.

	S ug/liter	V ug/liter/hr	T hr	K ug/liter	v ug/liter/hr	% respiration
Glu	0.78	19.00	1.7	31.5	0.46	52
Asp	1.56	27.84	1.5	40.0	1.04	54
Lys	0.62	2.47	10.3	24.9	0.06	24
Phe	0.53	1.47	19.8	28.5	0.03	25
Ala	2.90	13.91	2.3	27.4	1.11	43
Leu	0.92	2.01	3.5	6.1	0.26	20
Ile	0.73	1.48	4.0	5.2	0.18	22
Pro	0.79	2.71	6.6	17.2	0.12	44
Val	1.30	2.36	4.6	10.0	0.28	28
Tyr	1.11	7.61	26.3	198.9	0.04	33
Thr	2.70	3.56	4.4	12.9	0.62	36

terial activity. In the example (Fig. 9), the V is expressed as per gram dry weight of sediment and because of this is impossible to compare with plankton measures. The turnover times can be compared and the activity appears to be from 10 to 50 times greater in the sediment than in the plankton. In most of the samples, the highest activity was at the surface of the sediments and little activity was noted below 23 cm.

When the isotope uptake method is tried in the ocean, it is only successful in areas of relatively high production. For example, Vaccaro et al. (1968) made a transect across the Atlantic and found usable uptake kinetics for glucose only in coastal waters. However, when samples were each incubated with glucose for a number of hours, the uptake pattern appeared. In the Pacific, the uptake patterns are also erratic and could not be correlated with other parameters measured by R. D. Hamilton (personal communication). However, with most samples from the open ocean, it is impossible to tell anything about the past history of that particular patch of water. Thus, the heterotrophic activity may be the result of an earlier photosynthetic burst (as in the Erken samples).

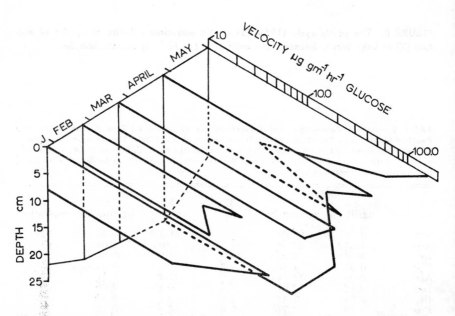

FIGURE 9. The maximum velocity of uptake (V) of glucose in the sediments of the Pamlico River, North Carolina (1969). Measurements made on a diluted sample and the data corrected back to the original concentration (L. W. Wood, unpublished data).

The final use of the isotope method is as a bioassay for the specific organic compounds. This was proposed by Hobbie and Wright (1965). We used a bacterial culture which was added to filtered lake and river waters. Previous tests had determined the K value for this culture in media where only A was present. Therefore, the (K + S) of the culture bacteria in the filtered water gave a measure of S by subtraction. By this method, glucose was found to be present at 5 to 20 μg/liter levels. Allen (1968) modified this by measuring the (K + S) first on a normal, untreated sample (obtained (K + S)), and then on a sample diluted 1 to 1 with artificial lake water (obtained (K + S/2)). Solving these two simultaneous equations yielded S. Tests of the bioassay against a chemical measure of glucose (Vaccaro et al., 1968) gave good agreement.

Conclusions

The radioisotope uptake method of studying heterotrophic bacteria is really a measure of the bacterial transport systems for various substrates. If the maximum velocity of uptake is measured and the concentration of substrate known, then the flux of organic carbon through the bacteria can be calculated. Unfortunately, the flux of only a few substrates has been measured, but enough is known to show that an important quantity of carbon is entering the bacteria. Because they are converting dissolved organic material into particulate material suitable for zooplankton to eat, the bacteria may be an important part of the food web.

These isotope studies have confirmed that there is a small but active portion of the dissolved organic carbon pool turning over every few hours. In polluted waters, the concentration of dissolved organic compounds is kept low by the rapid uptake into bacteria and so the flux rate must be known if the importance of various compounds is to be evaluated.

Either through growth or induction of new uptake sites, the heterotrophic bacteria are able to respond to increases in the available organic compounds. In eutrophic environments the bacteria are always active and the response is a strong seasonal cycle tied to the algal and detrital cycles. In oligotrophic environments, such as the open ocean, the bacteria appear to be in some resting state most of the time. However, if incubated for a day with organic substrates they are able to induce transport mechanisms; this suggests that they are opportunists waiting for fresh detritus or a burst of photosynthesis to resume activity. Thus, the heterotrophic bacteria in the aquatic ecosystem are closely coupled to their environment and are so well adapted to taking up low concentrations of simple sugars, amino acids, and short-chain fatty acids that they keep these compounds at low concentrations. There have even been suggestions that the coupling may be so close as to involve daily cycles in tune with photosynthetic excretion of dissolved organic material.

Literature Cited

Allen, H. L. 1968. Acetate in fresh water: Natural substrate concentrations determined by dilution bioassay. Ecology 49:346-349.

Allen, H. L. 1967. Acetate utilization by heterotrophic bacteria in a pond. In Hungarian symposium, Problems of organic matter determination in freshwaters. Hidrologia i Közlöny 47:295-297.

Alexander, M. 1961. Introduction to soil microbiology. J. Wiley and Sons, Inc., New York.

Hobbie, J. E. 1967. Glucose and acetate in freshwater: concentrations and turnover rates, p. 245-251. In H. L. Golterman and R. S. Clymo (eds.), Chemical environment in the aquatic habitat. Amsterdam.

Hobbie, J. E. and R. T. Wright. 1965. Bioassay with bacterial uptake kinetics: glucose in freshwater. Limnol. Oceanogr. 10:471-474.

Hobbie, J. E., C. C. Crawford, and K. L. Webb. 1968. Amino acid flux in an estuary. Science 159:1463-1464.

Hobbie, J. E. and C. C. Crawford. 1969. Respiration corrections for bacterial uptake of dissolved organic compounds in natural waters. Limnol. Oceanog. 14:528-532.

Kuznetsov, S. I. 1968. Recent studies on the role of microorganisms in the cycling of substances in lakes. Limnol. Oceanogr. 13:211-224.

Nauwerck, A. 1963. Die Beziehungen zwischen Zooplankton und Phytoplankton im See Erken. Symb. Bot. Upsaliens. 17(5):1-163.

Rodhe, W., J. E. Hobbie, and R. T. Wright. 1968. Phototrophy and heterotrophy in high mountain lakes. Verh. Internat. Verein. Limnol. 16:302-313.

Thayer, G. 1969. Phytoplankton production and factors influencing production in the shallow estuaries near Beaufort, North Carolina. Unpublished Ph.D. thesis, North Carolina State University, Raleigh.

Vaccaro, R. F., S. E. Hicks, H. W. Jannasch, and F. G. Carey. 1968. The occurrence and role of glucose in seawater. Limnol. Oceanogr. 13:356-360.

Watt, W. D. and G. E. Fogg. 1966. The kinetics of extracellular glycollate production by Chlorella pyrenoidosa. J. Exper. Botany, 17:117-134.

Wetzel, R. 1969. Excretion of dissolved organic compounds by aquatic macrophytes. BioSci. 19:539-540.

Wright, R. T. In press. Glycollic acid uptake by planktonic bacteria. In Proceedings of the symposium on organic matter in natural waters, September 2-4, College, Alaska.

Wright, R. T. and J. E. Hobbie. 1966. Use of glucose and acetate by bacteria and algae in aquatic ecosystems. Ecology. 47:447-464.

Bruce C. Parker and Mary Ann Wachtel

Seasonal Distribution of Cobalamins, Biotin and Niacin in Rainwater

Abstract

Bioassays of rainwater for dissolved cobalamins, biotin, and niacin during 15 months in the St. Louis area show appreciable quantities of these vitamins may reach terrestrial and aquatic ecosystems via atmospheric precipitation. At least for biotin and niacin, the frequency of occurrence and concentration of vitamins is higher during the growing season relative to the November 1 to March 31 period. Limited evidence suggests airborne soil particles and pollen as sources of these vitamins. In addition, clouds are conceived as hypothetical atmospheric ecosystems containing metabolically active microorganisms associated with cloud condensation nuclei and nutrient-rich water droplets.

Introduction

Parker (1968) first reported the occurrence of cobalamins (Vitamin B_{12}) in St. Louis rainwater. This discovery occurred during preliminary tests in a goldfish pond of a new sampling apparatus called a biodialystat (Parker, 1967a, b). On several occasions, the B_{12} concentration in biodialystats rose significantly 12 to 24 hours following initiation of spring rains, and this rise in B_{12} coincided with a bloom of *Chlamydomonas* in the pond. A few days later, the B_{12} concentration dropped concurrently with decline in the algal bloom.

Parker (1968) confirmed that the bulk of cobalamins added to the pond came from rains, rather than runoff, vegetation drip, or *in situ* production. B_{12} was so concentrated in several of these 1967 rains derived from convective storms that one could well imagine an appreciable impact of rainborne cobalamins on other aquatic ecosystems in the St. Louis area. Indeed, if a hypothetical lake containing *Euglena gracilis* and a sufficiency of all nutrients except B_{12} received a rainfall of 1.0 cm containing 20 pg/ml cobalamins, *Euglena* could increase to about 10^6 cells/cm^2 of lake surface, according to Parker's calculation. This would produce a visible bloom. The papers of Provasoli (1958), and Menzel and Spaeth (1963) also bear out the probability that cobalamins may limit natural phytoplankton populations.

Results of this earlier study stimulated the initiation of research to evaluate the seasonal contribution of rainfall in the St. Louis area to the cobalamin reserve and cycle in lakes. This program soon expanded to include biotin and niacin.

Methods

Soluble B_{12} was assayed with *Euglena gracilis* Z strain using the general procedure of Robbins, Hervey, & Stebbins (1950). Biotin and niacin assays utilized the bacterium *Lactobacillus Plantarum*, according to the methods outlined in Difco Manual, 9th ed. (1953).

Rainwater was collected both at the senior author's home and at Washington University in open stainless steel or aluminum containers placed on the roof away from overhanging vegetation. Most collections were frozen immediately following collection and prior to filtration through 0.22-pored GS Millipore filters. A total of 106 collections were made between April 25, 1967 and June 9, 1969. Initially collections were sporadic, but from June 1968 to June 1969 almost every rain and snow in excess of 0.1 cm liquid depth was collected for assay. The total rainfall collected during this last year approximates 87 cm, which approaches the average annual rainfall measured by the U. S. Weather Bureau in this region. Therefore, we are confident that our collections have been fairly complete.

Euglena assays for cobalamins utilized hand counts with hemacytometer during the first year and Coulter electronic particle counter subsequently. The inoculum was standardized at 600 cells/ml. Assays were run in triplicate, usually with reasonably low variation among replicate flasks. Standard curves based on *Euglena* growth for known concentrations of cyanocobalamin varied only slightly by our method. As performed in our laboratory, we consider the assay for B_{12} reliable for concentrations above 0.20 pg/ml.

Lactobacillus assays for biotin and niacin utilized turbidity measurements at 540 mμ on a B & L Spectronic 20 colorimeter. Except where noted, results are expressed as means of two separate assay runs, each consisting of two cultures (i.e., four replicates). We find greater variation with replicate cultures of these bacteria. Thus, by our method, we consider only levels above 1.0 pg biotin and 1.0 mμg niacin as reliably measured.

We have used several specific modifications in the published methods in order to improve the reliability of our assays:

(1) All Pyrex glassware is pretreated in a muffle furnace 30 minutes at 600°C to remove trace organics adhering to the glass.

(2) In biotin and niacin assays, we omit the saline solution step, which

is a dilution to achieve less dense inocula. Instead we substitute an extra transfer step using vitamin-deficient medium.

(3) Also, in assaying biotin and niacin turbidimetrically, we read the turbidity of the first tubes in a series both at beginning and end. Because the reading of a series of 20 duplicate tubes may take 15 minutes, these log-phase bacterial cultures frequently show a significant change in turbidity in 15 minutes.

(4) Prior to media-sterilization, the autoclave is scrubbed with 95% ethanol. We have shown at least for vitamin B_{12} (Parker, 1968) that residual vitamin derived from prior autoclaving of complex media will contaminate vitamin-free media at these high heats and pressures.

Results

Table 1 summarizes the data, including that reported earlier for cobalamins (Parker, 1968). The term "nil" is used here both for values that are negative or too low for accurate detection. Such negative values were especially common with cobalamin assays. A range in concentrations for biotin or niacin indicates that one experiment gave the lower reading while the other gave the higher.

Note the numerous instances where a series of collections was made during continuous rainfall. Of 15 series collections, cobalamins varied appreciably during the course of the rain only in one case; on July 15, 1968 the first fraction of rain contained 0.20, while the second fraction contained 4.50 pg cobalamin/ml. Of 12 series collections, biotin exhibited a downward trend in concentration during the course of rainfall in six cases, and no appreciable change for the other cases. Of 13 series collections, niacin showed some variability during the course of rains; however, the downward trend in concentration occurred in at least several experiments. In general, the concentrations of vitamins exhibit no consistent trend during long-term rains. Furthermore, these data provide only weak support for the idea of an early rain-out of vitamins, which might be predicted if the vitamins were derived from airborne dust. Also, in several attempts to correlate B vitamin concentrations with total dry weight of non-filterable particulates in rains, we have found no consistent pattern.

Three rains (April 17, May 7-8, and May 15, 1968) were collected in a large volume of ethanol to preclude any vitamin contribution from microbial activity between the beginning of rain collection and freezing. For assay, the ethanol was removed by vacuum distillation at reduced temperature prior to filtering. Note that April 17, 1968 rain contained

TABLE 1.—Concentrations of Cobalamin, Biotin, and Niacin in Rainfall, St. Louis, Missouri, April 25, 1967 to June 9, 1969.

Collection Date	Vitamin concentration/ml (means of all replicates)			Calculated Vitamin/cm² of Land and/or Water Surface (conc/ml X cm rainfall)		
	Cobalamin (pg)	Biotin (pg)	Niacin (mμg)	Cobalamin (pg)	Biotin (pg)	Niacin (mμg)
4–25–67	20.00			15.20		
6–21–67	0.25			0.78		
6–27–67	20.00			32.00		
7–12–67	nil			nil		
8–15, 16–67	nil			nil		
1–29, 30–68	nil			nil		
3–18–68	3.80	0–5.0	1.4	1.10	0–1.5	0.4
4–17–68†	9.80			1.10		
4–28, 29–68	nil			nil		
5– 7, 8–68†	nil			nil		
5–15–68†	nil			nil		
5–23–68	nil			nil		
5–21–68ᵃ	nil			nil		
5–22, 23–68ᵇ	nil			nil		
5–23–68ᶜ	nil			nil		
5–23, 24–68ᵈ	nil			nil		
5–28, 29–68	nil	3.5	2.9	nil	?	?
5–31–68ᵃ	nil			nil		
5–31–68ᵇ	nil			nil		
5–31, 6–1–68ᶜ	nil			nil		
6–11–68	<0.20			<0.20		
6–15–68	<0.20	4.3	0–1.0	<0.30	6.5	0–1.5
6–16–68	<0.20			<0.02		
6–19–68	1.40			<0.14		
6–22–68	5.60	9.1	0–4.4	2.80	4.6	0–0.7
6–25–68	<0.20	6.5	0–1.2	<0.14	4.6	0–0.8
7– 1–68	<0.20			<0.18		
7–14–68	0.20	3.4	0–1.0	0.75	12.8	0–15.0
7–15–68ᵃ	<0.20			0.04		
7–15–68ᵇ	4.50			?		
7–17–68	<0.20	5.8	nil	<0.48	13.9	nil
7–18–68	0.30	5.0*	nil	0.05	0.8	nil
7–24–68ᵃ	0.20	9.1	2.4*	0.06	2.7	0.7
7–24, 25–68ᵇ	<0.20	5.0	nil	<0.06	1.5	nil
7–25–68ᶜ	<0.20	5.5	3.2	<0.20	5.5	3.2
7–25–68ᵈ	<0.20	2.7	nil	<0.03	0.4	nil
7–26–68	<0.20	5.4	nil	<0.04	1.1	nil
7–31–68	<0.20	5.9	0–1.2	<0.08	2.4	0–0.5
8– 7–68	0.20	19.0*	3.2*	0.05	4.8	0.8
8–14–68	<0.20	7.5	2.2	<0.04	1.5	0.4
8–15–68	0.20	4.2	0–3.8	0.20	4.2	0–3.8
8–17–68	<0.20	6.0	1.8	<0.04	1.2	0.4
8–30, 31–68	<0.20	18.0	3.8	<0.12	10.8	2.3
9–15, 16–68	<0.20	0–3.6	0–4.4	<0.10	0–1.8	0–2.2
9–17–68	0.20	13.2	0–4.0	0.14	9.2	0–2.8
9–17, 18–68 (unfrozen)	<0.20	0–1.0	0–3.6	<0.54	0–2.7	0–9.7
9–17, 18–68 (frozen)	<0.20	0.2.0	0–4.0	<0.54	0–5.4	0–10.8
9–19–68	<0.20	3.0	3.0	<0.16	2.4	2.4
9–24–68	<0.20	3.3	3.0	<0.10	1.7	1.5
9–28, 29–68		3.5	2.9			
10– 5, 6–68	<0.20	13.0	3.6	0.06	3.9	1.11
10–13, 14–68	<0.20	3.8	3.0	<0.20	3.8	3.0
10–17–68	4.10	18.0	4.6	2.05	9.0	2.3
11– 2, 3–68ᵃ	<0.20	7.4	0–0.6	<0.20	7.4	0–0.6
11– 4–68ᵇ	<0.20	6.2	3.3	<0.40	12.4	6.6

TABLE 1.—**Concentrations of Cobalamin, Biotin, and Niacin in Rainfall, St. Louis, Missouri, April 25, 1967 to June 9, 1969.—Continued.**

Collection Date	Vitamin concentration/ml (means of all replicates)			Calculated Vitamin/cm^2 of Land and/or Water Surface (conc/ml X cm rainfall)		
	Cobalamin (pg)	Biotin (pg)	Niacin (mμg)	Cobalamin (pg)	Biotin (pg)	Niacin (mμg)
11- 7–68	<0.20	0–5.0	0–1.6	<0.20	0–6.5	0–1.6
11–14, 15–68[a]	<0.20	0–4.4	nil	<0.20	0–4.4	nil
11–15–68[b]	<0.20	nil	nil	<0.92	nil	nil
11–17–68[a]	.....	....	0–0.8	?	?	?
11–17, 18–68[b]	<0.20	0–2.0	nil	<0.10	0–1.0	nil
11–23–68	0.27	9.0*	1.7	0.03	0.9	0.2
11–26, 27–68[a]	<0.20	7.6*	3.6*	<0.16	6.08	2.9
11–28–68[b]	<0.20	2.0	0–1.0	<0.24	2.4	0–1.2
11–30, 12–1–68	<0.20	0–0.4	0–0.8	<0.28	0–0.6	0–1.1
12–18–68[a]	0.30	0–18.0	nil	0.30	0–18.0	nil
12–18–68[b]	0.88	nil*	2.0*	0.13	nil	0.3
12–21, 22–68	<0.20	nil	0–1.4	<0.16	nil	0–1.1
12–27–68[a]	<0.20	nil	0–0.4	<0.64	nil	1.3
12–27, 28–68[b]	<0.20	nil	0–0.4	<0.22	nil	0–0.4
12–30–68	<0.20	nil	0–1.0	<0.20	nil	0–1.0
1–16–69[a]	<0.20	nil	0–0.6	<0.22	nil	0–0.7
1–17–69[b]	<0.20	nil	0.8	<0.04	nil	0.2
1–17–69[c]	<0.20	nil	nil	<0.30	nil	nil
1–21–69	<0.20	nil	nil*	<0.03	nil	nil
1–22–69	<0.20	nil	nil	<0.18	nil	nil
1–23–69	<0.20	0–1.6	nil	?	?	nil
1–27–69[a]	<0.20	nil	nil	<0.10	nil	nil
1–28–69[b]	<0.20	nil	nil	<0.22	nil	nil
1–29–69[c]	<0.20	nil	nil	<0.42	nil	nil
1–29–69[d]	<0.20	nil	nil	<0.10	nil	nil
2- 5, 6–69	0.30	1.1	0–1.6	0.09	0.3	0–0.5
2- 7, 8–69	1.50	0–0.4	nil	3.30	0–0.9	nil
2–16, 17–69 (snow)	8.00	2.1	1.9	?	?	?
2–22–69	<0.20	2.1	nil	<0.28	2.9	nil
2–27, 28–69	nil	0–0.2	nil	nil	0–0.4	nil
3- 7–69	nil	2.0	1.5	nil	2.2	1.7
3–23, 24–69	nil	1.7	2.1	nil	5.6	6.9
3–26–69	nil	2.3		nil	0.7	
4- 4–69 (rain)[a]		2.1	2.4		2.1	2.4
4- 4–69 (thunber)[b]	nil	1.6	2.6	nil	2.2	3.6
4- 8, 9–69	nil	0–1.6	1.7	nil	0–4.2	4.4
4–13–69[a]	nil	nil	1.9	nil	nil	1.9
4–14–69[b]	0.20	nil	1.8	?	nil	?
4–14–69[c]	nil	0–1.2	1.0	?	?	?
4–17, 18–69[a]	0.22	2.0	2.8	0.22	2.0	2.8
4–18–69[b]	nil	1.6	1.9	nil	1.4	1.7
4–19–69	nil	2.5	3.6	nil	1.0	1.4
4–27–69	0.22	2.5	3.6	?	?	?
5- 7, 8–69	0.20	6.9	14.9	0.08	2.8	6.0
5–13–69	0.20	4.7	4.0	0.28	6.6	5.6
5–21, 22–69	<0.20	6.8	3.3	<0.20	6.8	3.3
5–31–69	<0.20	2.2	1.7	<0.22	2.4	1.9
5–31, 6- 1–69	<0.20	nil	2.2	<0.14	nil	1.5
6- 1–69	nil	2.4	2.1	nil	10.8	9.5
6- 8–69[a]	<0.20	2.1	1.6	<0.90	9.5	7.2
6- 8, 9–69[b]	0.23	2.5	2.0	0.18	2.0	1.6

*One experiment only; insufficient water for repeat.
†Collected in 95% EtOH, later removed by vacuum distillation before assay.
a, b, etc. refers to consecutive collections, regarded here as a single rain.
?indicates inability to calculate due to inaccurate rainfall depth measurement.

appreciable cobalamin. We have not repeated this ethanol-collection procedure for biotin and niacin.

In Table 1 rain occurring on September 17-18, 1968 was divided into two portions after collection. One fraction was frozen in routine fashion, the other was left unfrozen and prepared immediately for bioassay. Although the results of replicate assays of this rainwater showed some variation, there was no indication that freezing prior to filtration caused appreciable change in assayable vitamins.

As noted previously, our experience with the assays for these vitamins suggests that concentrations above 0.20 pg cobalamin/ml, 1.0 pg biotin/ml, and 1.0 mμg niacin/ml are reliably detected. Table 2 lists those rains which contained one or more vitamins in this detectable range. This table makes it obvious that a significant level of biotin or niacin occurred more frequently than cobalamin. Generally, the lower levels for all three vitamins occurred during winter months. The only high value during winter for cobalamin was 8.00 pg/ml, derived from a heavy snow storm on December 16-17, 1969. Also of interest, the highest values for biotin and niacin during 1969 occurred on May 7-8, the period when the major quantity of pine pollen was released in the vicinity of the collection vessels. This pollen fell with the rain in significant visible amounts.

Table 2 shows that all three vitamins occurred in significant amounts in only 5 rains, cobalamins and biotin occurred together 6 times, and cobalamins occurred with niacin 7 times. In contrast to these relationships, biotin and niacin occurred in significant amounts concurrently 28 times.

Table 3 presents additional calculations derived from Table 1. These calculations show that biotin and niacin occurred in significant amounts in 78 and 61% of the rains during the growing season, respectively. Such frequencies of occurrence are considerably higher than the value of 19% for cobalamins. Table 3 also shows that the average concentrations of vitamins in rain, especially during the growing season, are well above the levels at which the growth of assay organisms are stimulated. The estimated annual contribution of these vitamins by rains in the St. Louis area are at best conservative.

Discussion

This research, while leaving many questions unanswered, has proven that rainwater may contribute significant amounts of water-soluble vitamins to terrestrial and aquatic ecosystems. The concept of rain as a vector for inorganics must now be expanded to include trace organic substances. Such substances may act in the aquatic environment as (1) energy sub-

TABLE 2.—Rains With Over 0.20pg Cobalamin, 1.0pg Biotin, and 1.0mμg Niacin per ml.

Collection Date	Cobalamin	Biotin	Niacin
1967:			
4–25	20.00		
6–21	0.25		
6–27	20.00		
1968:			
3–18	3.80		1.4
4–17	9.80		
5–28, 29		3.5	2.9
6–15		4.3	
6–19	1.40		
6–22	5.60	9.1	
6–25		6.5	
7–14		3.4	
7–15	4.50		
7–17		5.8	
7–18	0.30		
7–24[a]		9.1	
7–24, 25[b]		5.0	
7–25[c]		5.5	3.2
7–25[d]		2.7	
7–26		5.4	
7–31		5.9	
8–14		7.5	2.2
8–15		4.2	
8–17		6.0	1.8
8–30, 31		18.0	3.8
9–17		13.2	
9–19		3.0	3.0
9–24		3.3	3.0
9–28, 29		3.5	2.9
10–5, 6		13.0	3.6
10–13, 14		3.8	3.0
10–17	4.10	18.0	4.6
11–2, 3[a]		7.4	
11–4[b]		6.2	3.3
11–23	0.27		1.7
11–28		2.0	
12–18[a]	0.30		
12–18[b]	0.88		
1969:			
2–5, 6	0.30	1.1	
2–7, 8	1.50		
2–16, 17	8.00	2.1	1.9
2–22		2.1	
3–7		2.0	1.5
3–23, 24		1.7	2.1
3–26		2.3	
4–4[a]		2.1	2.4
4–4[b]		1.6	2.6
4–8, 9			1.7
4–13[a]			1.9
4–14[b]			1.8
4–17, 18[a]	0.22	2.0	2.8
4–18[b]		1.6	1.9
4–19[c]		2.5	3.6
4–27	0.22	2.5	3.6
5–7, 8		6.9	14.9

TABLE 2.—Rains With Over 0.20pg Cobalamin, 1.0pg Biotin, and 1.0m μg Niacin per ml.— Continued.

Collection Date	Cobalamin	Biotin	Niacin
5–13		4.7	4.0
5–21, 22		6.8	3.3
5–31, 6–1			2.2
5–31		2.2	1.7
6–1		2.4	2.1
6–8		2.1	1.6
6–8, 9	0.23	2.5	2.0

TABLE 3.—Calculation Summarizing Data from Table 1.

Calculations	Cobalamin	Biotin	Niacin
Total rain samples assayed	103	81	81
Rains with significant concentrations of vitamins*	19 (19%)	45 (55%)	34 (42%)
Rains assayed April 1 to November 1	68	46	46
Rains April 1 to November 1 with significant concentrations of vitamins*	12 (18%)	36 (78%)	28 (61%)
Rains assayed November 1 to March 31	35	35	35
Rains November 1 to March 31 with significant concentrations of vitamins*	7 (20%)	9 (26%)	6 (18%)
Average concentration of vitamin/rain, April 1 to November 1	0.98pg/ml	4.1pg/ml	2.0m μg/ml
Average concentration of vitamin/rain, November 1 to March 31	0.43pg/ml	0.75pg/ml	0.42m μg/ml
Estimated annual contribution of vitamins per cm² by St. Louis rains	74	250	108

*Significance here refers to 0.20pg cobalamin, 1.0pg biotin, and 1.0m μg niacin per ml, as justified in text.

strates, (2) carbon skeletons, (3) vitamins, (4) inhibitors, (5) chelators, etc. Parker (1968) reported that rains, in the St. Louis area at least, sometimes contained more than 8 mg/liter of dissolved organic matter, a concentration equal to or higher than that found in several small lakes in that region. Thus, rains may increase the concentration of total dissolved organics in lakes as well as modify the *in situ* organic matter composition. The level of 8 mg/l total dissolved organic matter in St. Louis rains is about 10^6 times the sum of concentrations of the three vitamins treated in this study. Other workers have detected such trace organics as chlorinated hydrocarbon pesticides and herbicides, terpenoids, organic oxidants, etc. (Went, et al., 1967; Parker and Barsom, 1970). Therefore, the main bulk of rainborne organic solutes is still unidentified.

In the St. Louis area, the concentrations of dissolved cobalamins in rainwater sometimes exceed that within lakes in that area. We have not done biotin and niacin assays on St. Louis lakes; however, the concentrations of these two vitamins in rainwater are orders of magnitude above the levels reported in lakes and the ocean (Vallentyne, 1957; Carlucci and Sibernagel, 1967). Consequently, we conclude on theoretical grounds that the contribution of these three vitamins by rain may augment and influence appreciably some aquatic environments.

Our data are too limited for an elaborate evaluation of the ecological importance of rainborne vitamins to aquatic ecosystems. First, the responses of assay organisms in axenic-defined media are not necessarily equitable with that of vitamin-requiring members of microbial communities. Even those several organisms used for assay of single vitamins have varying degrees of sensitivity and specificity for analogs (Baker and Sobotka, 1962; Carlucci and Sibernagel, 1967). Second, our data treat only that fraction of the vitamin content of rain that passes through the 0.22μ pores of GS Millipore membranes. So far, no method for determining sestonic or total vitamin has proven satisfactory in this laboratory (Parker, 1969). Even our measured filterable vitamin levels may be inaccurate. We know, for example, that 45 mm-diameter Millipore membranes absorb up to 25 pg of cobalamins, and this amount is not elutable with water. Thus, our procedure may cause loss of up to 0.25 pg cobalamins/ml by membrane absorption, and this phenomenon may account in part for the lower frequency of occurrence of significant B_{12} concentrations in rainwater. We have not determined the amounts of biotin and niacin absorption on Millipore membranes. Third, the rather high frequency of negative values for our *Euglena* B_{12} assay of rainwater may stem from co-existence of antimetabolites for this vitamin. Consequently, our bioassay results for these vitamins represent only rough approximations of the real concentrations dissolved in rainwater.

A more thorough evaluation of the quantitative importance of rainwater organics to aquatic systems necessitates more information on (1) other constituents of rainwater, (2) their frequency and quantity of distribution, (3) the sources of such organics, and (4) conditions relating to their occurrence. We have only scratched the surface on these subjects. Our preliminary data suggest little connection between the total particulates and the concentrations of these vitamins in rains. Also, several attempts to extract cobalamins from airborne dust have given *nil* values, suggesting that either the extraction method failed or dry airborne dust is not a major source of this vitamin. For example, calcareous clays absorb B$_{12}$ tenaciously, and their presence in airborne dust in the St. Louis area may interfere with cobalamin extraction.

A clue to one possible source of B vitamins in rains comes from the correlation between pine pollen and high biotin and niacin levels during the 1969 spring. Pollen and spores often comprise a major fraction of airborne dust and rainwater particulates (McDonald, 1962; Pop, et al., 1964; Gregory and Monteith, 1967). Also the available pollen and spore count data for the St. Louis area suggest that the high values of niacin and biotin during the fall of 1968 coincide somewhat with peaks for weed pollen, while high niacin during late March-May, 1969 correlates with peaks for tree pollen. The dominant trees in the St. Louis area are *Pinus*, *Picea, Salix, Populus, Quercus*, and *Ulmus*, all of which produce appreciable airborne pollen during spring.

In conclusion, we wish to suggest a second possible origin for vitamins and possibly other organics in rain. We propose that part of these substances may be synthesized by microorganisms living within clouds. Once borne into the amtosphere, microorganisms can remain suspended for long periods; for example, a 50μ-sized particle will remain suspended indefinitely in the atmosphere over St. Louis if it is not further aggregated or brought down by rain.

Fischer et al. (1969) records that continental air may contain on the average 100 particles/cc ranging in diameter from 0.2-2.0μ. The values are lower, but nevertheless significant, for progressively larger particles up to about 50μ diameter. The composition of these large cloud-inhabiting particulates has not, to our knowledge, been elucidated. However, Gregory (personal communication) has confirmed the presence of microorganisms in clouds over Britain. Also the works of Schlichting (1961), Brown et al. (1964) for algae and that of Zobell (1942) for bacteria document that viable microorganisms from the seas as well as land surface can reach high altitudes in the atmosphere.

According to our hypothesis, cloud microorganisms and organic particulates comprise a small fraction of the condensation nuclei for rain droplets. During the period of condensation, which may vary up to many days, extensive microbial activity within unfrozen condensation droplets of some clouds occurs. The diameters of cloud droplets measure up to 100μ, and we have no reason to suspect such droplets are nutrient-poor on the basis of airborne dust and salt nuclei (Aufm Kampe and Weickmann, 1957; Weickmann, 1957). Furthermore, cloud moisture should absorb some of the harmful radiation which might otherwise kill these microorganisms or destroy the vitamins produced.

This hypothesis does not exclude contributions of organic matter to rain from other sources. Indeed Woodcock (1955) noted that, in addition to "giant" hygroscopic nuclei arising from bursting air bubbles at the sea surface, a sizeable amount of organic matter also entered the atmosphere from sea. Also, the works of Valencia (1967) and Garrett (1968) show that large amounts of non-volatile organic matter from the sea surface is borne into the atmosphere (see also Parker and Barson, 1970). Wilson (1959) found bacteria, niatoms, fragments of phyto- and zoo-plankton borne into the atmosphere over New Zealand from bursting air bubbles. Neumann (1959) also proposed that the large amounts of organic matter found in the atmosphere might also arise from the sea surface.

Our preliminary observations of airborne dust indicate that some microorganisms (yeasts, algae, bacteria) may remain vegetative and metabolically active within the air. May and Druett (1968) also have demonstrated viability of select microorganisms on threads which presumably simulate airborne moisture conditions. If microorganisms can undergo metabolism within clouds, then these and their organic products may augment the composition of rainwater derived from other mechanisms. Frequently, freezing in clouds at these high altitudes just prior to rainfall, a feature especially common to thunder clouds (Weickmann, 1957), may squeeze soluble vitamins from cells during falling rains.

To our knowledge, no one previously has envisioned clouds as living ecosystems. Direct evidence supporting this hypothesis is lacking at this time, as is also the evidence to refute it. We hope to test this hypothesis in the course of examining further the influence of rainborne organic substances on aquatic ecosystems.

Acknowledgments

We are grateful to the Center for the Biology of Natural Systems, Washington University, for support of this research under PHS grant, P10 ES 00139; to Kay Williams for assistance with some of the vitamin assays.

Literature Cited

Aufm Kampe, H. J. and H. K. Weickmann. 1957. Physics of clouds. Meteorol. Res. Rev. 3:182-225.

Baker, H. and H. Sobotka. 1962. Microbiological assay methods for vitamins. Adv. Clin. Chem. 5:173-235.

Brown, R. M., D. A. Larson and H. C. Bold. 1964. Air-borne algae: Their abundance and heterogeneity. Science 143:583-5.

Carlucci, A. F. and S. B. Sibernagel. 1967. Bioassay of seawater. IV. The determination of dissolved biotin in seawater using ^{14}C uptake by cells of *Amphidinium carteri*. Canad. J. Microbiol. 13:979-86.

Difco Manual (authors anonymous). 1953. Difco manual of dehydrated culture media and reagents for microbiological and clinical laboratory procedures.

Fischer, W. H., J. P. Lodge, Jr., J. B. Pate and R. D. Cadle. 1969. Antarctic atmospheric chemistry: Preliminary exploration. Science 164:66-7.

Garrett, W. D. 1968. The influence of monomolecular surface films on the production of condensation nuclei from bubbled seawater. J. Geophys. Res. 73:5145-50.

Gregory, P. H. and J. L. Monteith (eds.) 1967. Air-borne microbes. A symposium of the Society for General Microbiology (London), April, 1967. Cambridge University Press, N. Y. 1-397.

McDonald, J. E. 1962. Collection and washout of air-borne pollens and spores by raindrops. Science 135:435-6.

May, K. R. and H. A. Druett. 1968. A microthread technique for studying the viability of microbes in a simulated air-borne state. J. Gen. Microbiol. 51:353-66.

Mensel, D. W. and J. P. Spaeth. 1962. Occurrence of vitamin B$_{12}$ in the Sargasso Sea. Limnol. and Oceanogr. 7:151-4.

Neumann, G. H., S. Fonselius and L. Wahlman. 1959. Measurements on the content of non-volatile organic material in atmospheric precipitation. Int. J. Air Pollut. 2:132-41.

Parker, B. C. 1967a. Biodialystat: New sampler for dissolved organic matter. Limnol. and Oceanogr. 12:722-3.

——————. 1967b. Influence of method for removal of seston on the dissolved organic matter. J. Phycol. 3:166-73.

——————. 1968. Rain as a source of vitamin B$_{12}$. Nature. 219:617-8.

——————. 1969. Influence of method for removal of seston on the dissolved organic matter. II Cobalamins. J. Phycol. 5:124-7.

—————— and G. Barsom. 1970. Biological and chemical significance of surface microlayers in aquatic ecosystems. Bioscience 20:87-93.

Pop, E., N. Boscain, R. Flavia, B. Diaconeasa and A. Todoran. 1964. Effects of atmospheric precipitations on the pollen and spores concentration from the aeroplankton. Rev. Roumaine Biol. Ser. Bot. 9:329-34.

Provasoli, L. 1958. Nutrition and ecology of protozoa and algae. Ann. Rev. Microbiol. 12:279-308.

Robbins, W. J., A. Hervey and M. E. Stebbins. 1950. Studies of *Euglena* and vitamin B$_{12}$. Bull. Torrey Bot. Club. 77:423-41.

Schlichting, H. E. 1961. Viable species of algae and protozoa in the atmosphere. Lloydia 24:81-8.

Vallencia, M. J. 1967. Recycling of pollen from an air-water interface. Amer. J. Sci. 265:843-7.

Vallentyne, J. R. 1957. The molecular nature of organic matter in lakes and oceans with lesser reference to sewage and terrestrial soils. J. Fish. Res. Bd. Canada 14: 33-82.

Weickmann, H. K. 1957. Physics of precipitation. Meteoral. Res. Rev. 3:226-55.

Went, F. W., D. B. Slemmons and H. N. Mozingo. 1967. The organic nature of atmospheric condensation nuclei. Proc. Nat. Acad. Sci. 58:69-74.

Wilson, A. T. 1959. Surface of the ocean as a source of air-borne nitrogenous material and other plant nutrients. Nature 104:99-101.

Woodcock, A. H. 1955. Bursting bubbles and air pollution. Sewage Industr. Wastes 27:1189-92.

Zobell, C. E. 1942. Microorganisms in marine air. Contribution No. 157, Scripps Inst. Oceanogr.

Robert A. Paterson

Lacustrine Fungal Communities

Abstract

Although representatives of all fungi are to be found in the aquatic habitat, the most abundant in terms of numbers of species are the aquatic Phycomycetes. In planktonic and benthic communities of the lacustrine ecosystem chytridiaceous and saprolegniaceous fungi have significant roles as parasites of algae and decomposers of bottom detritus. In the case of epidemic fungal infections of planktonic algae, activities of fungi may affect the composition of phytoplankton communities by delaying the time of algal maximum and by reducing the population of certain algae so that other phytoplankters will replace the infected algal populations. In the case of other infections which are not epidemic the fungi may not influence populations of algae during periods of maximum algal population. Instead, the fungi may only infect phytoplankters during periods of decline in the algal population and thus only hasten the decomposition of the plants. Some chytridiaceous fungi grow only on algal cells which are obviously dead. The foregoing indicates that there are probably three situations with regard to the relationship of these fungi to algae in the plankton: (1) fungi may be obligately parasitic attacking living algal cells during periods of active growth, (2) fungi may not be obligately parasitic and only attack algal cells in a senescent condition during periods of decline in the algal population, and (3) the fungus lives only on dead algal cells.

There appears to be a pattern to the distribution of chytridiaceous planktonic fungi that infect diatoms at various depths in a lake. For the most part the oldest structures in the fungal life cycles are found at the greatest depths, indicating that fungus infections may occur near the surface and that the life cycles of the fungi proceed as the diatom cells drop to the bottom.

Aquatic Phycomycetes are commonly found in the benthic community. Fungi most commonly found during studies of Douglas Lake, Michigan, were members of the Chytridiales that decompose chitin. Thus these organisms may play a significant role in chitin decomposition in this lake. However, studies in Lake Michigan where the bottom depths were greater indicated the oomycetous fungi were the most common.

Introduction

The role of green plants as primary producers in the aquatic environment has received considerable attention. In addition, the contribution of bacteria to the nutritional cycles in aquatic ecosystems has become better understood in the last few years. Probably least known is the role of the true fungi in the freshwater habitat. Although representatives of all major groups of fungi have been found in aquatic situations, the most abundant in terms of numbers of species are undoubtedly the Phycomycetes. In the lacustrine ecosystem more is probably known about the Phycomycetes that are planktonic and benthic than those found elsewhere in lakes.

Discussion

Planktonic Fungi

The occurrence of phycomycetous thalli and such propagules as zoospores, gemmae, and hyphal fragments in the plankton has been well established. Indeed, many investigators have shown the presence of zoospores and other propagules in lake waters. Noteworthy among these are workers who have attempted quantitative or distributional studies of these structures. For example, Suzuki (1960, 1961b, c) and Suzuki and Nimura (1961) have attempted to enumerate the phycomycetous spores present in several Japanese lakes and to determine their distribution. Studies by Willoughby and Collins (1966; Collins and Willoughby, 1962) in lakes in England utilized laborious plating methods to determine zoospore numbers. Fuller and Poyton (1964) have isolated fresh water and marine Phycomycetes by plating techniques and special methods involving superspeed centrifugation. Although the latter method has not been fully tested it appears to be the most promising quantitative technique for determining numbers of zoospores and other propagules in lake waters. Such other methods as those used by the Japanese and English workers have not yielded reliable quantitative data.

Although the occurrence of phycomycetous propagules in lake waters is of considerable interest, many of these structures originate from thalli that are not planktonic. Indeed, many fungal sporangia that contribute zoospores to the plankton are from members of the Oomycetes and are undoubtedly found in the bottom sediments. Exceptions are species of such genera as *Leptolegnia*, *Aphanomyces*, and other fungi that attack planktonic crustaceans and rotifers (Paterson, 1958; Petersen, 1910; Prowse, 1954; Scott, 1961). Other exceptions are species that grow on what Sparrow (1968) terms adventitious plankters. Examples of these are dead insects, pollen and other floating debris.

Of the truly planktonic fungi only those members of the Chytridiales that attack phytoplankton have been studied quantitatively. Several years ago Canter and Lund (1948, 1951, 1953) investigated the effects of fungal infection on planktonic algal populations in English lakes. Their careful quantitative work revealed several interesting situations with regard to the relationship between fungal parasitism and factors acting on diatoms and on other algae. One situation was concerned with the effect of parasitism on algal maxima. Where the percent of parasitism was as high as 40 or more of the frustrules infected with chytridiaceous thalli there was a delay in the time of the algal maxima. In addition, the expected number of diatoms in a maximum were decreased. In another situation an increase in population occurred in uninfected algae when an algal species in the same community was parasitized by chytrids. As described by Canter and Lund from their study of Esthwaite water, the course of events that involved one blue-green alga and three species of diatoms was as follows: During the spring of 1949 the presence of *Oscillatoria agardhii* var. *isothrix* Skuja in great abundance (100 to 250 filaments per ml.) near the surface of the lake had the effect of reducing light penetration. The effect reduced the growth rate of *Asterionella formosa* Hass., *Tabellaria fenestrata* var. *asterionelloides* Grun. and *Fragilaria crotonensis* (Edw.) Kitton. In early May the *Oscillatoria* filaments became heavily infected by *Rhizophydium megarrhizum* Sparrow causing a rapid decrease in the numbers of *Oscillatoria* filaments. As the latter decreased there was a rapid increase in numbers of *Fragilaria* and *Tabellaria* colonies. During the last part of May *Asterionella formosa* became infected by *Rhizophydium planktonicum* Canter causing a rapid decline in the numbers of diatom colonies. The species of *Fragilaria* and *Tabellaria* remained uninfected and continued to increase in growth rate until the middle of June when silica became lacking and thus limited further increase in growth. A third observation by Canter and Lund indicated that the amount of infection of the algal population was related to the relative growth rates of the host and the fungus. Indeed, they reported infections so severe that the number of fungal thalli was greater than the number of algal cells.

In other studies Canter and Lund (1948) observed that occasionally an epidemic occurred when a host population was about to decline due to other causes. They suggested that the decrease in numbers of algae in a population was hastened by the parasitism. Koob (1966) and Paterson (1960) observed parasitism of phytoplankters by chytrids to occur only during periods of decline in the infected algal population. Further, there was no apparent hastening of the decline in numbers of algae. Indeed in the study by Koob of *Asterionella formosa* parasitized by *Rhizophydium planktonicum* in two Colorado lakes the fungus infection did not appear to have the effect observed by Canter and Lund. Although the parasitism reached a maximum of 40% during Koob's investigation he could find no evidence that parasitism hastened the decrease in numbers of diatom cells. In the study by Paterson in a Michigan lake the infestation of *Anabaena planktonica* Brunnthaler by *Rhizosiphon anabaenae* (Rodhe and Skuja) Canter also occurred during the decline in the algal population. However, the percent parasitism was very small and never exceeded 6% of the algal filaments. In this case the percent parasitism was too small to have any effect on the algal population. Therefore, Koob's and Paterson's investigations suggest that in some situations chytrids attack algae only during periods of algal decline and probably have little or no effect on the algal population.

It is evident from the foregoing that the seasonal distribution of chytrids infesting live algal cells is determined by the seasonal periodicity of the host. Obviously, the fungus can only be observed to infect the host when the host is present. However, at least one planktonic chytrid has been shown to have a periodicity in seasonal distribution when the host cells are continuously present. In a study by Paterson (1960) of the chytrid *Amphicypellus elegans* Ingold which infects only dead thecae of *Ceratium hirundinella* (O. F. M.) Schrank the chytrid was present only during June and August. The percent of *Ceratium* cells supporting the growth of the chytrid was high, being 38% in June and 92% in August. The host was present continuously from the first of June to the middle of October. Although *A. elegans* cannot be considered a parasite it is nevertheless planktonic.

The foregoing discussion on planktonic fungi indicates that the relationships between the planktonic chytrids and their algal hosts is not the same in each situation. I suggest that there are at least three relationships. In the first case the percent parasitism is high and the alga is attacked during its growth phase before a maximum in numbers of algae is reached. This suggests that the fungus is truly parasitic in that it attacks only actively growing, healthy algal cells. In the second situation algal cells are attacked only during the declining phase of the population after a maximum in their numbers is reached. This indicates that the fungus is specific in its requirements and different from the first case. It does not infest active living cells but only those becoming moribund or senescent. In the third situation the fungus attacks only dead cells or non-living parts of an alga. An example is the invasion of dead *C. hirundinella* thecae by *A. elegans*. In the first two situations the affinity of the fungus for the alga is probably specific. Although *A. elegans* has been found on dinoflagellates other than *Ceratium* it is probably specific for that group. It undoubtedly has a requirement for some substance in the thecal wall since the protoplasm is always observed to be absent in infested cells of the host and the rhizoids appear to be only in the wall.

Studies by Koob (1966) and an unpublished investigation by Paterson have shown two interesting features of chytrid specificity. Koob found in his populations of *Asterionella formosa* five distinct classes of colonies based on frustule length. Further, he found that only the populations of one of the size classes was infected by *Rhizophydium planktonicum*. The other four populations remained uninfected. Paterson also found in his investigation of *Chytridium marylandicum* Paterson a specificity within a host species. *C. marylandicum* attacks *Botryococcus braunii* Kütz. However, the fungal plant body does not penetrate the living cells of the colony, but is found only in the mucilaginous sheath. During the investigation *B. braunii* was isolated in unialgal cultures from each of seven lakes in Michigan and one in Maryland. The cultures were inoculated with *C. marylandicum*. In spite of repeated attempts at culturing and inoculation growth did not occur in the cultures of *B. braunii* isolated from three of the Michigan lakes. The fungus grew abundantly on the four other algal isolates from Michigan and on the one from Maryland. Therefore, there are some features of specificity that remain unknown at present.

Concerning the role of chytridiaceous fungi which infest phytoplankters in the lacustrine ecosystem Wesenburg-Lund (1905) stated the following. "I have shown, further, that nearly all the protoplasm of the cells in plankton is eaten away by Phycomycetes before reaching the bottom; my observations prove that an organism in the latter part of the period of maximum development may very often be infected by Phycomycetes, which feed upon the protoplasm and kill it leaving the skeleton intact." A study by Paterson (1967) on the vertical distribution of chytridiaceous fungi tends to support Wesenburg-Lund's contention that phytoplankters are attacked and sink with their parasites to the bottom of lakes. Table I shows the vertical distribution of three chytrids which were found in greatest abundance and their algal hosts on two days a week apart in Lake Michigan. Of the three chytrids, *Zygorhizidium melosirae* Canter occurred most frequently. It was found on July 16 at the surface when it infected *Asterionella formosa* at 15, 25, and 30 m and on July 22 at 2 m when it infected *Synedra* sp. Even though those thalli which occurred at 2 m on July 22, and at the surface on July 16 probably belonged to the same fungus population, it is possible that *Z. melosirae* at the other depths were members of different populations that could have been separated from the former by the thermal stratification of the lake. That thermal stratification can influence the distribution of the host, thereby limiting the invasion by chytrid zoospores, has been suggested

TABLE I.—Vertical Distribution of Some Fungi and Temperature Data from Grand Traverse Bay, Lake Michigan, 1963.

Temp (7–16)	Depth (meters)	7–16	7–22
19.5	0	*	
	2		
19.0	5		—
18.2	10		
16.5	11	≡	
11.5	15	s—	
8.5	20	<	
			<
7.8	25	≡	
7.2	30	s—	
		<	
	80		<

* = *Zygorhizidium melosirae* on *Asterionella formosa*
— = *Zygorhizidium melosirae* on *Synedra* sp.
≡ = *Rhizophydium fragilariae* on *Fragilaria crotonensis*
< = *Rhizophydium* sp. on *Tabellaria flocculosa*
s = Sexual stages

by Lund (1957). It seems entirely possible that a thermal stratification may also restrict the movement of zoospores from layer to layer. In this investigation a major discontinuity layer existed at 10 m and 11 m. There was thus a thermoclime between chytrids at the surface and the populations at 15, 25, and 30 m.

Two populations were probably involved in the distribution of *Rhizophydium fragilariae* Canter infecting *Fragilaria crotonensis* (A. M. Edw.) Kitton. The population of *R. fragilariae* at 10 m on July 16 and that at 15 m on July 22 were probably the same, the latter having sunk from 10 to 15 m during the week. This situation with regard to *Rhizophydium* sp. infecting *Tabellaria flocculosa* (Roth) Kütz is similar. The infested diatoms were found at the lower depths of 20 and 30 m on July 16, and 30 and 80 m on July 22. These distribution patterns were not due to the presence of the host only at these depths since they were generally present at all depths on both dates. It appears that the infected diatoms seem to have sunk in the water from July 16 to July 22.

Evidence that those fungal populations at the lower depths were older than those near the surface was the presence of sexual structures in Z. *melosirae* at 15 and 30 m but not at the surface and 2 m. According to Canter and Lund (1948) the sequence of events in a fungal infection is as follows. The zoospore cysts appear first on the host and are most abundant. Then sporangia with protoplasm are most numerous, followed by empty sporangia and then sexually-formed resting spores. Therefore, it is probable that populations of Z. *melosirae* at 15 and 30 m were relatively old. Perhaps they were formed previously nearer the surface and had sunk to those depths just previous to July 16.

Benthic Fungi

Although the presence of chytridiaceous and oomycetous fungi in lake bottoms has been well established, little else is known other than some information about their distribution and ecology. Most studies have been made using the typical phycomycetous substrates or "baits." Some examples are cellophane, chitin of insects, snake skin, hemp seeds, grass leaves, and pollen. In a study by Paterson (1967) in Douglas Lake, Michigan, "baits" were placed in containers that were put on the bottom of the lake at various depths from .5 m to 15 m. the most common fungi found were such chitinophilic chytridioceous species as *Asterophylctis sarcoptoides* H. E. Petersen, *Obelidium mucronatum* Nowakowskiella, *Rhizoclosmatium aurantiacum* Sparrow, *Chytridiomyces chyalinus* Karling and a species of *Siphonaria*. The greatest number of chytrid species were recovered from the bottom of Douglas Lake during the first week or 10 days in July to the end of that month, whereas very few were found in August. Although very few oomycetous fungi were recovered from the bottom of Douglas Lake they were found only during the first 2 weeks in July. Concerning distribution by depth, some species of chitrids were restricted to shallow waters, others to deep waters, and others were ubiquitous. Studies were also conducted by Paterson (1967) on Grand Traverse Bay, Lake Michigan. As in the investigation of Douglas Lake, substrates or "baits" were placed on the bottom. In Lake Michigan the areas studied were 26 and 31 m deep. Few chytrideaceous fungi were found and the most numerous fungi recovered were members of the Oomycetes. No seasonal distribution was evident in Grand Traverse Bay. Willoughby (1961a,b; 1962, 1965) has studied the distribution of fungi in the muds of English lakes. In these studies chitinophilic fungi and chytridiaceous species growing on cellulosic substrates were found abundantly. Saprolegniaceous propagules in Blelham Tarn were more frequently collected from soils of the lake margin than from the muds in the center of the lake. High numbers of mucors were collected from bottom muds indicating that fungous propagules sedimented and remained viable. However, these fungi could make no contribution to the aquatic ecosystem since they are terrestrial forms. Suzuki (1961a,d,e) has observed seasonal changes in the distribution of oomycetous fungi. More of these organisms were present in the bottom of Lake Nakanuma during the winter than during the summer. Fungi in other Japanese lakes had different seasonal distribution patterns. For example, in Lake Yamanakako the greatest numbers of fungi were found in the fall (Suzuki and Hatakeyama, 1961). The distribution of aquatic fungi in Japanese lake bottoms shows a relationship to the amount of dissolved oxygen. During periods of anaerobic conditions fungi in lake muds were scarce. For more complete information on Suzuki's work consult Sparrow (1968).

Conclusion

In conclusion, planktonic and benthic fungi are commonly found in lacustrine habitats. In some situations fungal parasitism may have an effect on increasing or decreasing populations of phytoplankters. In other situations chytridiaceous fungi may be decomposers that break down the algae as they become senescent or die and fall to the bottom of a lake. Benthic fungi are composed of chytridiaceous or oomycetous fungi and are found with enough frequency to conclude that they contribute to the breakdown of such substances as chitin, cellulose, and perhaps other materials as yet unknown.

Literature Cited

Canter, H. M., and J. W. G. Lund. 1948. Studies on plankton parasites. I. Fluctuations in the numbers of *Asterionella formosa* Hass. in relation to fungal epidemics. New Phytologist 47:238-261.

————. 1951. Studies on plankton parasites. III. Examples of the interaction between parasitism and other factors determining the growth of diatoms. Annals of Botany (London) New Series 15:359-371.

————. 1953. Studies on plankton parasites. II. The parasitism of diatoms with special reference to lakes in the English lake district. British Mycological Society Transactions 36: 13-37.

Collins, V. G., and L. G. Willoughby. 1962. The distribution of bacteria and fungal spores in Blelham Tarn with particular reference to an experimental overturn. Archiv fur Mikrobiologe 43:244-307.

Fuller, M. S. and R. O. Poyton. 1964. A new technique for the isolation of aquatic fungi. Bioscience 14:45-46.

Koob, D. D. 1966. Parasitism of *Asterionella formosa* Hass. by a chytrid in two lakes of the Rawah wild area of Colorado. Journal Phycology 2:41-45.

Lund, J. W. G. 1957. Fungal diseases of plankton algae, p. 19-23. *In* C. Horton-Smith (ed.). Aspects of the transmission of disease. Hafner Publishing Company, New York.

Paterson, R. A. 1958. Parasitic and saprophytic Phycomycetes which invade planktonic organisms. II. A new species of Dangeardia with notes on other lacustrine fungi. Mycologia 50:453-468.

—————. 1960. Infestation of chytridiaceous fungi on phytoplankton in relation to certain environmental factors. Ecology 41:416-424.

—————. 1967. Benthic and planktonic Phycomycetes from northern Michigan. Mycologia. 59:405-416.

Petersen, H. E. 1910. An account of Danish freshwater-Phycomycetes, with biological and systematical remarks. Annales Mycologici 8:494-560.

Prowse, G. A. 1954. *Aphanomyces daphniae* sp. nov., parasitic on *Daphnia hyalina*. British Mycological Society Transactions. 37:22-28.

Scott, W. W. 1961. A monograph of the genus *Aphanomyces*. Virginia Agricultural Experiment Station Technical Bulletin 151:1-95.

Sparrow, F. K., Jr. 1968. Ecology of freshwater fungi, p. 41-93. *In* G. C. Ainsworth and A. S. Sussman (ed.), The fungi, an advanced treatise. Academic Press, New York and London.

Suzuki, S. 1960. Microbiological studies of the lakes of Volcano Bandai. I. Ecological studies on aquatic Phycomycetes in the Goshikinuma Lake group. Japanese Journal of Ecology 10:172-176.

—————. 1961a. The seasonal changes of aquatic fungi in lake bottom of Lake Nakanuma. Botanical Magazine (Tokyo) 74:30-33.

—————. 1961b. The vertical distribution of the zoospores of aquatic fungi during the circulation and stagnation period. Botanical Magazine (Tokyo) 74:254-258.

—————. 1961c. Distribution of aquatic Phycomycetes in some inorganic acidotrophic lakes of Japan. Botanical Magazine (Tokyo) 74:317-320.

—————. 1961d. Ecological studies on aquatic fungi in the lakes of Volcano Nikko. Japanese Journal of Ecology 11:1-4.

—————. 1961e. Some considerations on anaerobic life of aquatic fungi in lake bottom. Japanese Journal of Ecology 11:219-221.

————— and T. Hatakeyama. 1961. Ecological studies on the aquatic fungi in Lake Yamanakako. Japanese Journal of Ecology 11:173-175.

Suzuki, S., and H. Nimura. 1961. The vertical distributions of fungi and bacteria in lake during the circulation and stagnation period. Japanese Mycological Society Transactions 7:115-117.

Wesenburg-Lund, C. 1905. A comparative study of the lakes of Scotland and Denmark. Proceedings of the Royal Society of Edinburg 25:401-480.

Willoughby, L. G. 1961a. The ecology of some lower fungi at Esthwaite Water. British Mycological Society Transactions 44:305-332.

——————. 1961b. Chitinophilic chytrids from lake muds. British Mycological Society Transactions 44:586-592.

——————. 1962. New species of *Nephrochytrium* from the English lake district. Nova Hedwigia 3:439-444.

——————. 1965. Some observations of the location of sites of fungal activity at Blelham Tarn. Hydrobiologia 25:352-356.

—————— and V. G. Collins. 1966. A study of the distribution of fungal spores and bacteria in Blelham Tarn and its associated streams. Nova Hedwigia 12:150-171.

John Cairns, Jr.

Factors Affecting the Number of Species in Fresh-Water Protozoan Communities

Abstract

Although protozoans are affected by changes in the chemical, physical, and biological environment in the same general ways as other organisms, their probable cosmopolitan distribution is shared with only a few other groups. If the distribution of species is as general as some evidence indicates, this coupled with the ability of many species to encyst and excyst at appropriate times provides an opportunity for interspecific relationships which are not possible for many other groups of organisms. Thus, species with similar environmental requirements might occur together anywhere in the world as frequently as appropriate ecological conditions materialize. Under these conditions complex relationships might develop which would produce a variety of interactions between species, including interdependence and mutual exclusion. Thus one might view an aggregation of protozoan species as either (1) a structured community with a complex series of interlocking cause-effect pathways, or (2) an assortment of species present because their arrival or excystment coincided with acceptable ecological conditions. In the former case the composition of the community would be determined primarily by a combination of internal and external factors; and in the latter case by random chance or probability (i.e., the "right" species being deposited, by wind or animal, etc., or already being present as a cyst in the right place at the right time). In order to explore these possibilities the following factors were considered (1) chemical-physical environment, (2) interspecific relationships, (3) substrate, (4) colonization-extinction rates, (5) changes in community composition through space and time, (6) perturbations which drastically reduce the number of species in a community. Available evidence indicates that both internal and external factors regulate the number of species present.

Introduction

In 1948 the Academy of Natural Sciences began a series of river surveys under the direction of Dr. Ruth Patrick. Most of us participating in these surveys were impressed by the comparatively constant number of species in many major groups of aquatic organisms despite considerable variation in the kinds of species that comprised these aggregations (Table 1). Perhaps the word constant should be avoided and one should ask why the differences are around 20-30% rather than two or three orders of magnitude. These results have been reported in a number of papers—Patrick (1949, 1961) are representative of this series. This relatively constant number of species was particularly surprising for protozoans because (1) of the continual succession of species quite common in fresh-water protozoan communities,* (2) of the strong possibility that most protozoan species have a cosmopolitan distribution and may be found wherever appropriate ecological conditions develop, (3) of the aerial and ground transport systems described by Brown *et al.* (1964), Schlichting (1961, 1964), Maguire and Belk (1967) and others which suggest that protozoans (and other micro-organisms) are being deposited all over the world nearly continually, (4) of the large number of species collectively tolerating a wide range of ecological conditions which are available to colonize various habitats, (5) of the continual and often abrupt changes in environmental conditions in a particular habitat. Given these conditions, a variation in number of species of several hundred percent on either side of a mean would not be surprising; yet such variation would be exceptional in an unpolluted temperate zone river or stream in North America. The purpose of this paper is to examine the factors affecting the number of species in fresh-water protozoan communities, and to attempt to explain the observed small variation in the number of protozoan species. For a review of general ecological parameters I suggest Noland and Gojdics (1967).

Discussion

Changes in River Protozoan Community Composition Through Space and Time

The word community when used for organisms found in rivers or streams usually means that group of aquatic organisms found in a particular area at a particular time. The organisms, however, particularly protozoans, usually form a continuous gradient through both time and space with few or no sharp lines of demarcation that enable one to perceive distinct aggregations of organisms that one might call "communities." Actually although there are considerable differences in the types of protozoan species found in different areas of the same stream (Fig. 1) at approximately the same time, there is no sharp boundary between the areas (Table 2). The percentage overlap of species found at a single

---

*Community is used to designate an aggregation of species at a particular point in space and time.

area through time is also comparatively small, since abrupt changes in weather may cause swift changes in species composition (Table 3). Due to these factors it is quite difficult to delimit a protozoan community in a flowing water system in either time or space. Presumably there are limitations on interactions, for example, while upstream protozoans may influence downstream protozoans the reverse seems quite unlikely. In addition, although water flow in a river system is not laminar it seems unlikely that a protozoan could have any great influence on other protozoans more than a few centimeters on either side of it at right angles to the direction of flow. In view of the shifting patterns of micro-turbulence any significant long-lasting effect upon downstream protozoans may be unlikely unless the upstream protozoans are fairly numerous in all parts of the stream. When one considers the half-life of many organic compounds, long-lasting indirect effects through time and space seem unlikely except possibly when massive blooms occur.

All of this merely suggests that "macro-level" interactions between protozoan species in river systems are the exception rather than the rule and that protozoan "communities" are difficult to delineate at macro-territorial levels. This still leaves the possibility of all of these being determined at "micro-levels," which is where one would logically expect them to occur. Unfortunately the requirements of teamwork on the Savannah and other river surveys did not permit this detailed an examination, since the primary objective was to determine the number of species present in a given area with some attention to numbers of individuals and kinds of species. This aspect will be discussed in detail later. However, it is important to note that an understanding of the spatial relationships between individual protozoans as well as between protozoans and other associated microbial species is essential. Except for a few papers such as Spoon and Burbanck (1967) we have not progressed much beyond the work of Picken (1937). One might expect, on the basis of structural differences alone, the behavioral patterns of *Chlamydomonas* (a phytomonad), *Vahlkampfia* (an amoeba), *Stylonychia* (a hypotrich), and *Paramecium* to be quite different. Presumably physical contact would be greatest between species with similar behavior and the number of contacts might influence population density of species already present as well as the colonizing success of invading species. Physiological or chemical interactions may also be influenced by spatial relationships so an understanding of them is a major key to the understanding of the structure and function of protozoan communities. Unfortunately our present knowledge of spatial relationships in complex fresh-water protozoan communities is so primitive it is virtually useless. The matter deserves much more attention than it has received.

TABLE 1.—Total Number of Systematic Entities for Each Study in Each River.

	Soft Water Rivers						Hard Water Rivers								
	Escambia	Savannah 54	Savannah 55	North Anna	White Clay	Flint	N. Fork Holston	Rock Creek	Ottawa 55	Ottawa 56	Potomac 56	Potomac 57	Mean All Rivers	Mean Soft Rivers	Mean Hard Rivers
Algae	77	105	101	98	73	79	63	65	76	58	105	103	84	89	78
Protozoa	38	61	40	58	56	51	—	86	—	48	85	68	59	51	72
Insects	29	58	51	61	57	—	83	48	59	61	89	99	63	51	73
Fish	39	19	35	21	20	13	21	24	18	28	18	29	24	25	23

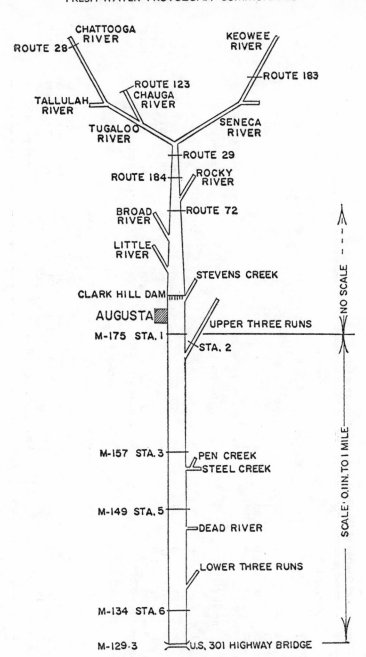

FIGURE 1. Diagrammatic map of Savannah River basin (from Patrick, Cairns, and Roback, 1967).

TABLE 2.—Percentage of Overlap of Species at Various Stations on a Survey.

Survey 1

Station	1	3	5	6
1	100	10	21	27
3	11	100	23	23
5	25	25	100	38
6	22	19	28	100

Average overlap—22.67

Survey 2

Station	1	3	5	6
1	100	63	45	50
3	50	100	47	42
5	32	41	100	45
6	29	30	35	100

Average overlap—42.42

Survey 3

Station	1	3	5	6
1	100	17	33	45
3	27	100	33	60
5	29	18	100	53
6	22	19	30	100

Average overlap—32.17

Survey 4

Station	1	3	5	6
1	100	50	37	60
3	40	100	35	53
5	34	43	100	43
6	30	35	23	100

Average overlap—40.25

Survey 5

Station	1	3	5	6
1	100	—	—	25
3	—	—	—	—
5	—	—	—	—
6	25	—	—	100

Average overlap—25.00

Survey 6

Station	1	3	5	6
1	100	42	42	45
3	33	100	38	31
5	37	40	100	51
6	22	30	45	100

Average overlap—36.75

Survey 7

Station	1	3	5	6
1	100	40	37	34
3	40	100	40	31
5	38	40	100	57
6	27	24	42	100

Average overlap—37.50

Survey 8

Station	1	3	5	6
1	100	21	25	25
3	22	100	27	30
5	22	22	100	29
6	24	28	31	100

Average overlap—25.50

Survey 9

Station	1	3	5	6
1	100	40	43	34
3	38	100	43	40
5	39	40	100	40
6	29	34	38	100

Average overlap—38.17

TABLE 3.—Percentage of Species Overlap Between Surveys on a Given Station.

Station 1

Survey	1	2	3	4	5	6	7	8	9
1	100	26	10	10	10	5	15	15	42
2	22	100	22	13	22	22	22	37	28
3	9	20	100	54	20	38	12	25	20
4	10	13	60	100	22	37	13	22	28
5	5	12	12	12	100	25	18	18	20
6	2	13	23	21	27	100	21	21	29
7	5	10	5	5	13	15	100	31	21
8	5	14	10	10	12	14	30	100	14
9	14	11	10	11	14	20	20	14	100

Average overlap—18.49

Station 3

Survey	1	2	3	4	5	6	7	8
1	100	18	—	18	5	18	18	30
2	10	100	10	25	25	29	14	32
3	—	20	100	67	33	27	27	20
4	10	25	35	100	29	21	18	14
5	1	13	10	15	100	13	18	20
6	8	21	10	15	19	100	19	27
7	7	9	9	10	19	14	100	29
8	9	15	5	7	18	18	24	100

Average overlap—17.86

Station 5

Survey	1	2	3	4	5	6	7	8
1	100	25	19	19	12	38	25	25
2	12	100	17	22	25	25	22	35
3	10	18	100	42	40	18	18	29
4	13	30	52	100	43	34	13	40
5	4	18	23	21	100	20	10	20
6	17	21	13	21	24	100	24	32
7	7	11	9	5	9	15	100	20
8	7	18	12	14	14	20	20	100

Average overlap—20.89

Station 6

Survey	1	2	3	4	5	6	7	8	9
1	100	28	13	22	13	19	31	40	13
2	15	100	15	25	23	23	18	23	12
3	7	12	100	34	20	17	20	22	20
4	11	23	40	100	27	20	27	23	14
5	5	17	19	20	100	30	22	29	27
6	9	19	17	17	32	100	22	17	29
7	12	12	19	20	22	20	100	39	24
8	13	13	17	15	22	12	31	100	17
9	5	10	19	11	25	25	24	20	100

Average overlap—20.11

Recently Cairns and Kaesler (1969) have used dendrograms to show similarity of communities through space and time (Fig. 2). This method enables one to quantify similarities and differences between aggregations of species.

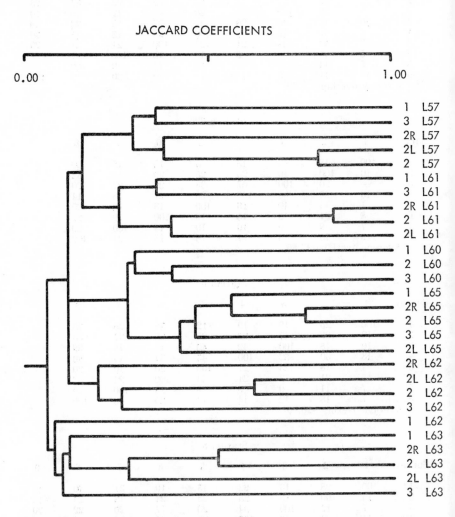

FIGURE 2. Dendrogram prepared by the unweighted pair-group method with arithmetic averages showing similarities among all low-water aggregations. The numbers to the right of the dendrogram indicate station number (i.e., sampling areas), the R or L after station 2 indicates right or left bank, the L associated with 57 through 65 indicates a low water survey in 1957, 1960, 1961, 1962, 1963, or 1965 (from Cairns and Kaesler, 1969).

Effects of Chemical-Physical Environment

One of the most critical factors in determining both diversity (number of species) and density of protozoan populations is water velocity. Zimmerman (1961) has shown that current velocities also affect the kinds of species found. In rivers the most diverse and dense protozoan populations developed in pond-like situations where the velocity of the current was not great (Cairns, 1965a). This was evident both from the comparison of stations with varying rates of flow and from the comparison of different flow rates within a single station. Laboratory experiments with natural lake water flowing through plastic troughs (Cairns and Yongue, 1968) suggest that there is an optimal flow rate which produces the greatest number of species. Although these experiments were not designed to determine the effects of flow rate, the average number of species was lowest in the system with the greatest flow, next highest in that with the least flow, and highest in those with an intermediate flow. It is possible that this represents a balance between a current too swift to permit certain species to become established and one too slow to bring sufficient nutrients past a given point to sustain other species. However, considerably better experimental evidence than this is needed. Probably it will be necessary to determine the velocity to which each individual protozoan is exposed, a fairly difficult task.

Determining the number of protozoan species present in a sample is affected by their density. There is a threshold density below which finding even a single specimen of a species becomes difficult. Large species are easier to locate than small ones so there is a bias in their favor. Perhaps this is not entirely bad since biomass is often important. However, probably only a fraction of the low density species are ever reported, and this may prove to be a major weakness in the data just discussed. Some of the difficulties in concentrating specimens and other sampling difficulties are discussed at length in Cairns (1965b).

Most of the chemical-physical data on environments in which protozoans have been found is too fragmentary to provide useful information on the range of conditions tolerated by most species. Even when a large amount of data is available it is not usually from the micro-habitat in which the protozoan was found but rather from a nearby area which may or may not have been the same. An example of this is given on p. 229 in Noland and Gojdics (1967). Probably the latter is the rule rather than the exception. Furthermore, the most critical environmental conditions influencing the kinds of species present may have been those which occurred some time before the sample was taken. However, the greatest obstacle is the low frequency of occurrence typical of most species. Even in a fairly extensive survey such as that of the Savannah River (Patrick, Cairns, Roback, 1967) most species were reported only once, or at best, only a few times, hardly sufficient evidence to provide acceptable ranges of tolerance. Those species found frequently in another study had a range of environmental conditions approximating the range of most North American temperate zone streams (Cairns, 1964). Although nearly twenty different types of determinations were included in this study, many important parameters (vitamin B-12, carbohydrate, etc.) were omitted and were probably major determinants for many species (for example, Chu, 1942; Rodhe, 1948; Hutner *et al.* 1949 and more recently Parker, 1968). At present we can only assume that most protozoan species are similar to, and possibly more sensitive than, the comparatively few species that have been carefully studied (since many of these may be "weed" species), and that most species have sufficiently complex requirements to insure that optimal conditions would be rare, suboptimal conditions more common, acceptable conditions frequent, and inadequate conditions common. A diagrammatic sketch of this model is given in Fig. 3.

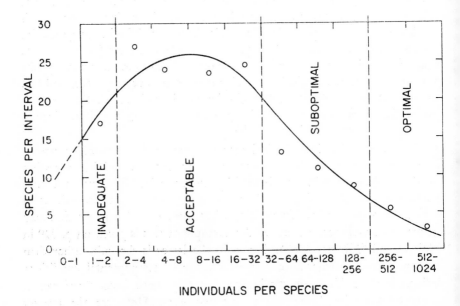

FIGURE 3. A model for the truncate normal curve distribution.

Species—Species Relationships

Many protozoan species collected today are remarkably similar morphologically to those described in the early days of light microscopy. Even if we discount these descriptions as inadequate, the similarity to descriptions of specimens examined by Kahl and other microscopists in the early part of this century is impressive. Apparently many species have been morphologically stable for many generations. Coupled with a cosmopolitan distribution, various transport systems, and the common ability to encyst and excyst, this apparent stability would appear to provide an opportunity for interspecific relationships to develop. In other words, if species have similar environmental requirements and a cosmopolitan distribution it might reasonably follow that the probability of interspecific relationships developing would be high. This possibility was explored by Cairns (1965b). In this study the frequency of occurrence of nearly 1200 protozoan species for 202 areas of various rivers and streams mostly in the United States was determined. Approximately 75% of the species occurred in three or less of the areas sampled, or less than about 1.6% of the time. The possibility of relationships developing between species which occur so infrequently seemed to be less likely than between species with a high frequency, so this possibility was examined with those which occurred four or more times in the 202 areas sampled. Of these, only 20, or roughly 6%, of those found four or more times occurred in at least 25% of the areas studied. The number of areas in which each of these 20 species occurred was as follows: 125, 87, 80, 69, 67, 66, 64, 64, 62, 60, 60, 60, 55, 53, 52, 49, 49, 48, 47, 47. Naturally these ubiquitous species occurred together quite often. In order to determine whether or not pairs of these species occurred together more frequently than would be expected from chance alone, an association matrix was made for the 20 sampling areas where these species were most common. A Chi-square test of significance was run on the 190 possible associations of species pairs, and of these, 44 occurred together more frequently than expected from chance alone at the 5% level of confidence. An examination of the environmental ranges for these species (Cairns, 1964), indicated: (1) pairs or larger groups of associated species always had virtually identical ranges of environmental conditions; (2) these species always tolerated rather broad ranges of environmental conditions; and (3) having identical ranges of tolerance to environmental conditions does not insure that species will be associated more often than would happen by chance alone, since only 44 pairs out of a possible 190 were associated significantly beyond chance despite the fact that all species had broadly overlapping environmental ranges.

In seeming contradiction to these results are those that indicate that some protozoans modify their environment in a manner beneficial to others, (for example, Robertson, 1921 and 1924; Picken, 1937; Mast and Pace, 1938; Kidder, 1941; Lily and Stillwell, 1962; Stillwell, 1967; Cairns, 1967) and that some species have very specific food habits (for example, *Didinium nasutum* rarely eats anything but *Paramecium* or *Actinobolina* anything but *Halteria*). Cairns (1967) speculated that if beneficial interactions existed one might expect natural selection to have shaped these interactions into synergistic associations or patterns of species. Once such a synergistic pattern developed for a given type of habitat, it would be advantageous for a prospective colonizer to be able to take advantage of that pattern, to be able to fit into it. To be associated with another pattern would be of less use to it, since successful invasion would then depend on the simultaneous invasion of several species. Thus, as a consequence of the high probability that protozoans may have a cosmopolitan distribution, we would expect that once such a synergistic pattern developed for a given type of habitat it would tend to become nearly universal for that type of habitat, and that the pattern would have acquired the property that each role or niche in it could be occupied by *any one of a number of species*. The species found to occupy that niche in a particular community at a given time would depend on some combination of environmental conditions and historical accident. This model would account for the relative constancy in number of species, the low frequency of association between particular species, and the successional pattern that is characteristic of most fresh-water protozoan communities. Unfortunately there is little evidence to support or reject this model. One of the major obstacles is that species are determined morphologically rather than functionally. It is possible that morphologically similar species may have dissimilar functions, while two morphologically dissimilar species may have nearly identical functions. However, if the model just described is valid then the relatively constant number of morphological species observed suggests a significant correlation between form and function. The synergistic pattern hypothesis has several attractive features and appears to be worth testing despite the considerable difficulties involved.

Substrate

In periodic studies of the Potomac River from 1956 to 1963 Cairns (1966) reports that although the types and even the amounts of various types of substrate remained remarkably constant, variations in kinds of species present as well as in the total diversity of species were common.

This suggests that these changes are initiated by environmental factors other than substrate quality. However, there is little doubt that unstable substrates, such as shifting sand, are less favorable habitats than algal mats growing on rocks. River substrates might thus be divided into two groups: (1) unfavorable—shifting sand, rocky wave-splashed shorelines, turbulent areas with swift currents, etc., and (2) favorable—algal mats, flocculent mud surfaces, mild currents, etc. In unpolluted situations the former usually have few individuals and few species, the latter, many species with a wide range of individuals per species. The substrate does not appear to directly determine the nature of succession or the kinds of species present unless some material is exuded into the environment. Cairns and Smith (unpublished manuscript) suspended various substrates, such as pine needles and apples, in the well-mixed epilimnion of Douglas Lake and found that some of these were colonized by markedly different protozoan communities from the others, although essentially the same kinds of species became established on a series of identical polyurethane substrates suspended in the same area.

Cairns and Yongue (1968) found that on a presumably homogeneous substrate (a plastic surface) the individuals of various protozoan species were not uniformly distributed. No patterns were evident to suggest a cause for the aggregations observed. One of the major weaknesses in analyzing data of this sort is the lack of information about the size of the area over which individual substrate associated protozoans move. No doubt this varies from species to species and is influenced by density of individuals of the same species, predator species, and prey species as well as by other factors. Studies of micro-distributional patterns and movements are difficult to make, because sampling may alter patterns and direct observation requires lighting sufficiently intense to influence results. Nevertheless information of this sort is badly needed.

Cairns and Ruthven (1970) suspended two groups of ten polyurethane substrates of different sizes in the well-mixed epilimnion of Dougas Lake, Michigan. The number of protozoan species in samples taken from the substrates over a six-week period was determined and comparisons were made between the number of species associated with substrates of various sizes, and between the number of species for each substrate at different sampling periods. The number of species associated with each substrate increased over time until a peak was reached at about 4-6 weeks; after that, for the substrates that peaked early, the number of species remained nearly constant or decreased slightly. Unfortunately an eight-week summer session which must include set up and close down time restricted the number of weeks of observation. Experiments now in progress with substrates installed about one month before the eight-week session began support the estimate of the leveling-off time. Results indicate that a minimal size range volume (area was not used because these substrates were porous) affects the number of protozoan species. Smaller substrates generally had fewer species than the larger substrates and a linear relationship was shown between log volume and the number of species (Figs. 4, 5). There were several sudden increases in number of species at various times for different substrates for which no cause was apparent. Some of the smaller substrates had disproportunately large numbers of species in the beginning of the experiment. It may be that optimal substrate size for a fresh-water protozoan community is partly a function of the *types* of species present. Since the arrival of appropriate colonizing species may be largely a matter of chance, one might then expect certain aggregations of colonizing species to be better suited to small substrates than others.

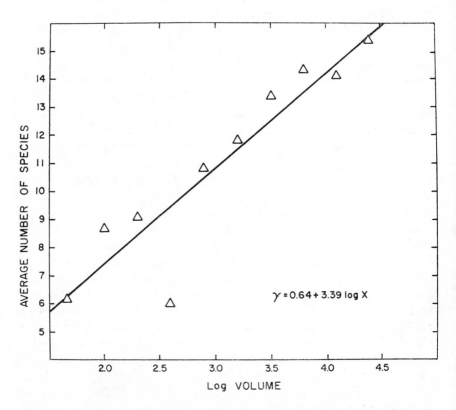

FIGURE 4. Relation between number of species of fresh-water protozoans and log substrate size (from Cairns and Ruthven, 1970).

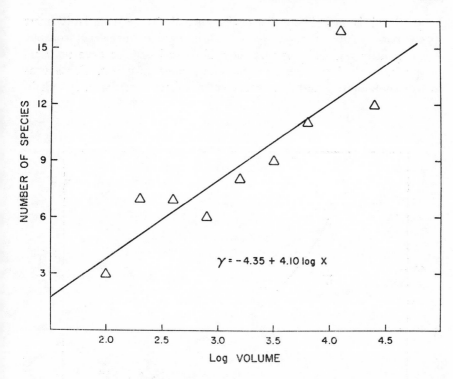

FIGURE 5. Relation between number of species of fresh-water protozoans and log substrate size (from Cairns and Ruthven, 1970).

Colonization—Extinction Curves

MacArthur and Wilson (1963, 1967) have proposed that islands with a continuing colonization of new species and a continuing extinction of established species may reach an equilibrium point between these processes which results in a relatively constant number of species being present despite succession. Cairns et al. (1969) studied two series of poly-urethane substrates (10 substrates/series) of identical size suspended in Douglas Lake, Michigan. Although the protozoan communities coloniz-ing each of the substrates were not identical, the colonization process was remarkably similar for the entire twenty substrates. In the early stages of invasion of a new substrate, the colonization rate greatly ex-ceeded the extinction rate as proposed by MacArthur and Wilson (Figs. 6, 7). However, as the number of species on each substrate increased, extinction and colonization rates approached but did not quite reach equilibrium. Equilibrium might have been reached had the summer ses-sion at the University of Michigan Biological Station been a few weeks longer, but it is more likely that environmental changes constantly upset species relationships so that a rigid equilibrium point is not very probable. A more likely possibility is that most biological systems have numbers of species oscillating about an equilibrium point or points. For a protozoan community one might conceive of this as a feedback system in

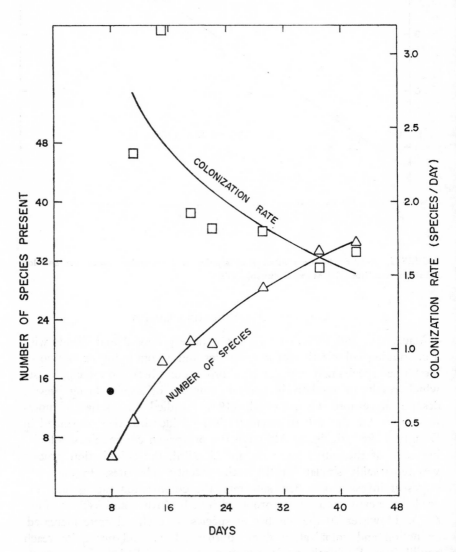

FIGURE 6. Colonization rate vs number of protozoan species on artificial substrates. (from Cairns et al., 1969).

which changes in the rate of one process inevitably affect the other, causing the number of species to vary constantly but within a specific range (barring some major and unusual environmental stress).

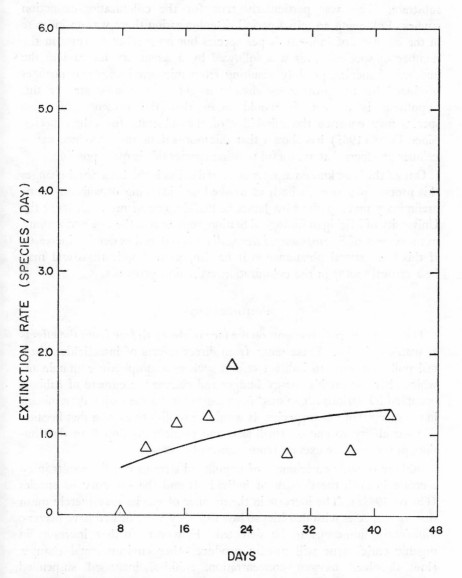

FIGURE 7. The extinction rate of protozoans on artificial substrates as a function of time in days (from Cairns et al., 1969).

In both the colonization-extinction studies (Cairns et al. 1969) and the substrate volume-number of species studies (Cairns and Ruthven, 1970) sudden increases in the number of species occurred (Figs. 8, 9). These did not appear to be correlated to external environmental events but rather to events associated intimately with the colonization of the substrate. This was particularly true for the colonization-extinction studies. Following an initial period of colonization there was an increase in the numbers of individuals per species but no marked increase in the number of species. This was followed by a secondary increase in the number of species, possibly resulting from microenvironmental changes produced by the protozoans already inhabiting the substrate. If this hypothesis is correct, it would mean that the presence of some species may enhance the suitability of the substrate for other species. Since Fogg (1965) has shown that microorganisms may produce extra-cellular products that are useful to other species this is quite possible.

One of the least known aspects of colonization is the behavioral changes this process produces in both established and invading organisms. Some preliminary investigations by James L. Plafkin, one of my students at the University of Michigan Biological Station, suggest that the rate and amount of movement of *Spirostomum intermedium* is reduced as density increases. If this is a general phenomenon it has important implications and may be a critical factor in the colonization-extinction processes.

Perturbations

There are few environments on the face of the earth free from the effects of man's activities. These range from direct effects of insecticides, thermal pollution, etc., to indirect effects such as atmospheric contaminants which alter the earth's energy budget and change the climate of habitats occupied by various organisms. Man may soon be the major determinant in the survival of many species. It would be foolish to assume that because of their ability to encyst, small size, and possible cosmopolitan distribution, protozoans are exempt from man-made stresses.

Initially organic enrichment of unpolluted streams usually results in an increase in both the density of individuals and the diversity of species (Cairns, 1965a). The increase in the number of species may merely mean that some species with densities so low that they were missed have become sufficiently numerous to be detected. However, further increases in organic enrichment will usually produce other environmental changes (low dissolved oxygen concentration, sulfides, increased suspended solids) unfavorable to many of the species present, which will eliminate many of them thus reducing the total number of species although the

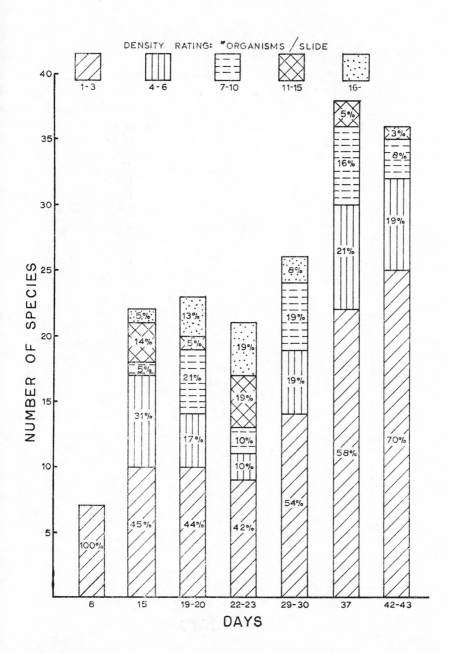

FIGURE 8. Number and percentage of protozoan species in each of five density-rating groups on artificial substrate at sampling location 5 (from Cairns et al., 1969).

JOHN CAIRNS, JR.

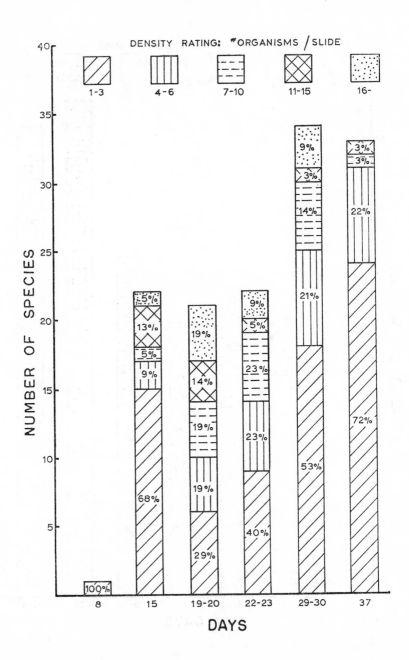

FIGURE 9. Number and percentage of protozoan species in each of five density-rating groups on artificial substrate at sampling loaction 8 (from Cairns et al., 1969).

numbers of individuals of tolerant species may actually increase (Cairns, 1965a). This simplification of a complex community (i.e., many species to few species) is the typical response of other major groups of aquatic organisms to environmental stress or pollution (Patrick, 1949). These responses were noted in aquatic communities in rivers and streams of the Conestoga Basin, Pennsylvania.

Similar responses (i.e., reduction in number of species) to high and low pH shocks as well as temperature shocks were noted in plastic troughs kept indoors with controlled photoperiod but with natural water from Douglas Lake flowing through them (Figs. 10, 11, and 12). These shocks lasted only a few minutes and since the original environmental conditions were quickly restored once the stress ceased the fairly rapid recovery rates are not surprising. Note that the number of species at the end of the recovery period approximates that found before the shock was applied and was also approximately that of the control (which often varied a few species + or —).

Non-selective, periodic, removal of protozoans from polyurethane substrates suspended in Douglas Lake resulted in an increased number of species when these substrates were compared to undisturbed controls exposed at the same time to similar environmental conditions (Cairns, Dickson and Yongue, 1971). This was accomplished by suspending 56 substrates in the epilimnion of Douglas Lake and squeezing certain selected substrates to collect a sample. These substrates were then reexamined after various time intervals and compared to control substrates which had had no previous examinations (and therefore no squeezing, which reduced the number of protozoans present when a sample was taken). The substrates that had periodic removal of portions of the protozoan community had higher numbers of species than those left undisturbed. Of course collecting a comparison sample removed them from the undisturbed class. When the latter substrates were reexamined after a sample had been taken the number of species had generally increased. These results indicate that potential colonizers may be excluded from an established protozoan community and that periodic removal of portions of the community favors establishment of new species. Since these substrates were suspended in Douglas Lake for a much longer period of time than substrates in previous experiments it would appear that these experiments were terminated approximately when peak species diversity was reached, and that an oscillation in number of species follows even when portions of the community are removed periodically. When the system reached a certain level of complexity or species diversity is oscillated near this level for the remaining period of observation (about six weeks for most substrates). Certain substrates sometimes had rather dramatic changes in number of species present—at times nearly 30 per cent. During this period the usual successional replacement of species occurred. Results also indicate that undisturbed substrates (those not previously sampled) first increase in number of species present and then the number of species present declines and oscillates about a mean that is significantly lower than the mean for disturbed substrates (Table 4). Whether either of these should be called "stable communities" is questionable but each seems to oscillate within a range of ± six species about 75% of the time.

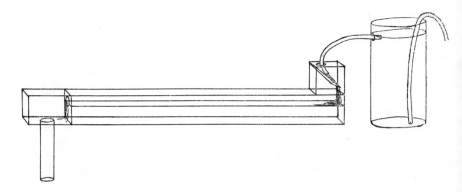

FIGURE 10. Plastic reservoir and trough used in temperature stress experiments (from Cairns, 1969).

Conclusions

Fresh-water protozoan communities appear to have one or more internal or endogenous regulatory systems that influence the number of species present at a particular time. There are two fairly strong sets of data supporting this statement: 1. The sudden increases in numbers of species that appear to be correlated to events associated intimately with the colonization of the substrate rather than to external environmental events (Cairns et al., 1969; Cairns and Ruthven, 1970). 2. The colonization of "undisturbed" substrates that was characterized by an increase in species followed by a decrease in species which rarely reached initial levels of diversity a second time. Oscillation of numbers of species on these undisturbed substrates was within a range significantly below that of "disturbed" substrates, suggesting that undisturbed communities excluded invading new species more effectively than disturbed communities (Cairns, Dickson and Yongue, 1970). On the other hand external events or exogenous factors seem to determine the number of species

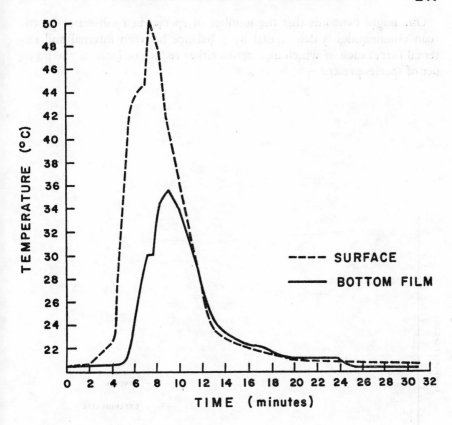

FIGURE 11. Pattern of thermal shock from about 20° to 50° C. Note that the water flowing over the surface film has a much higher temperature than the bottom film which contains most of the protozoan species (from Cairns, 1969).

present or the range within which oscillation in number of species will occur. There are several types of evidence supporting this statement: (1) the range within which the number of species oscillated was significantly higher on "disturbed" substrates than on "undisturbed" substrates. (2) The volume or size of substrate available for colonization as well as its physical structure influences the number of species present. (3) Sudden application of environmental stress (pH, temperature, zinc, copper) reduces the number of species present; the number of species usually returns to approximately its previous level when the stress is removed (Cairns, 1969, and Cairns and Dickson, 1970).

One might conclude that the number of species in fresh-water protozoan communities is determined by a balance between internal and external forces each of which may act to either reduce or increase the number of species present.

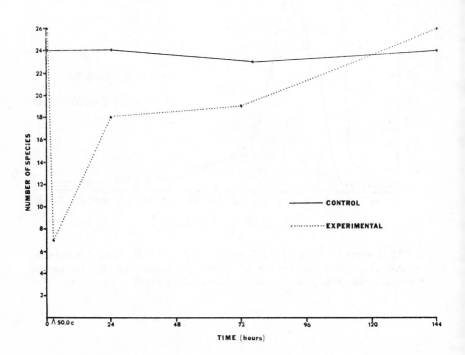

FIGURE 12. Changes in protozoan species diversity following the thermal shock in
Figure 11.

TABLE 4.—Number of Protozoan Species Found on Each Substrate for Each Sampling Period.

Substrate Series Number	Substrate Number in Series	Number of Times Sampled							
		1	2	3	4	5	6	7	8
46 Days in Lake									
I	1	28	44	39	40	29	31	31	54
	2	27	34	27	27	36	29	29	37
II	1	23	29	19	21	31	31	27	32
	2	20	23	30	18	29	31	24	35
III	1	18	27	18	17	24	35	—	—
	2	24	23	15	19	27	32	31	33
53 Days in Lake									
I	3	14	15	31	40	42	35		
	4	12	15	22	22	42	33		
II	3	15	25	20	25	31	35		
	4	13	21	19	21	19	22		
III	3	16	23	28	24	26	31		
	4	19	18	37	30	37	36		
60 Days in Lake									
I	5	34	28	28	*	31			
	6	22	22	26	34	31			
II	5	21	30	31	29	31			
	6	16	20	25	—	—			
III	5	21	26	26	25	32			
	6	14	21	27	27	27			
67 Days in Lake									
I	7	16	—	—	—				
	8	12	23	37	37				
II	7	14	19	—	—				
	8	16	21	27	29				
III	7	23	29	30	27				
	8	24	26	26	31				

TABLE 4.—Number of Protozoan Species Found on Each Substrate for Each Sampling Period.—Continued.

Substrate Series Number	Substrate Number in Series	Number of Times Sampled							
		1	2	3	4	5	6	7	8
74 Days in Lake									
I	9	19	24	24					
	10	18	23	—					
II	9	26	30	30					
	10	16	25	24					
III	9	20	32	21					
	10	17	31	30					
81 Days in Lake									
I	11	21	30						
	12	26	—						
II	11	19	27						
	12	15	29						
III	11	21	28						
	12	24	—						
88 Days in Lake									
I	13	22	24						
	14	25	37						
II	13	17	28						
	14	25	31						
III	13	23	29						
	14	33	35						
95 Days in Lake									
I	15	24							
	16	—							
II	15	21							
	16	12							
III	15	13							
	16	14							

Literature Cited

Brown, R. Malcolm, Jr., Donald A. Larson, Harold C. Bold. 1964. Airborne algae: their abundance and heterogeneity. Science, Vol. 143, pp. 583-585.

Cairns, John, Jr. 1964. The chemical environment of common fresh-water Protozoa. Not. Nat. Acad. Nat. Sci. Philadelphia, No. 365, pp. 1-16.

——————. 1965a. The Protozoa of the Conestoga Basin. Not. Nat. Acad. Nat. Sci. Philadelphia, No. 375, pp. 1-14.

——————. 1965b. The environmental requirements of Protozoa. Biological Problems in Water Pollution, Third Seminar, 1962, PHS Publ. No. 999-WP-25, pp. 48-52, Abst. pp. 385-386.

——————. 1966. The Protozoa of the Potomac River from Point of Rocks to White's Ferry. Not. Nat. Acad. Nat. Sci. Philadelphia., No. 387, pp. 1-11 plus 43 pp. supporting data deposited as document No. 8902 with the AID Aux. Pub. Proj. Photodupl. Serv., Library of Congress, microfilm copies $2.50.

——————. 1967. Probable existence of synergistic interactions among different species of Protozoans. Revista de Biologia, Vol. 6, Nos. 1-2, pp. 103-108.

——————, and W. H. Yongue, Jr. 1968. The distribution of fresh-water Protozoa on a relatively homogeneous substrate. Hydrobiologia, Vol. 31, No. 1, pp. 65-72.

——————. 1969. Rate of species diversity restoration following stress in protozoan communities. Univ. Kansas Sci. Bull., Vol. 48, No. 6, pp. 209-224.

——————, M. L. Dahlberg, K. L. Dickson, N. Smith, and W. T. Waller. 1969. The relationship of fresh-water protozoan communities to the MacArthur-Wilson equilibrium model. Amer. Nat., Vol. 103, pp. 439-454.

——————, and Roger L. Kaesler. 1969. Cluster analysis of Potomac River survey stations based on protozoan presence-absence data. Hydrobiologia, Vol. 34, No. 3-4, pp. 414-432.

——————, and Jeanne Ruthven. 1970. The relation between artificial substrate area and the number of fresh-water protozoan species. Trans. Amer. Micro. Soc., Vol. 89, No. 1, pp. 100-109.

——————, and Kenneth L. Dickson. 1970. Reduction and restoration of the number of fresh-water protozoan species following acute exposure to copper and zinc. Trans. Kansas Acad. Sci., Vol. 73, No. 1, pp. 1-10.

——————, Kenneth L. Dickson, and William H. Yongue, Jr. 1970. The consequences of nonselective periodic removal of portions of fresh-water protozoan communities. Trans. Amer. Micro Soc., Vol. 90, No. 1, pp. 71-80.

——————, and Nancy Smith (in preparation). The influence of substrate quality upon the colonization of fresh-water protozoans.

Chu, S. P. 1942. The influence of the mineral composition of the medium on the growth of planktonic algae. Jour. Ecol., Vol. 30, pp. 284-325.

Hutner, S. H., Provasoli, L., Stokstad, E. L. R., Hoffman, C. E., Belt, M., Franklin, A. L., and Jukes, T. H. 1949. Assay of antipernicious anemia factor with Euglena. Proc. Soc. Exp. Biol. Med., Vol. 70, pp. 118-120.

Kidder, G. W. 1941. Growth studies on ciliates. The acceleration and inhibition of ciliates grown in biochemically conditioned medium. Physiol. Zool., Vol. 14, pp. 209-226.

Lily, D. M. and R. H. Stillwell. 1965. Probiotics: Growth-promoting factors produced by microorganisms. Science, Vol. 147, pp. 747-748.

MacArthur, R., and E. O. Wilson. 1963. An equilibrium theory of insular zoogeography. Evolution, Vol. 17, pp. 373-387.

——————. 1967. The theory of island biogeography. Princeton University Press, Princeton, New Jersey.

Maguire, Bassett, Jr., and Denton Belk. 1967. Paramecium transport by land snails. Jour. Proto., Vol. 14, No. 3, pp. 445-447.

Mast, E. C., and D. M. Pace. 1938. The effect of substances produced by Chilomonas paramecium on the rate of reproduction. Physiol. Zool., Vol. 11, pp. 359-382.

Noland, Lowell E. and Mary Gojdics. 1967. Ecology of free-living Protozoa. Research in Protozoology, Vol. 2, Pergamon, pp. 217-266.

Parker, Bruce. 1968. The vitamin B-12 content of rainwater. Nature, Vol. 219, No. 5154, pp. 617-618.

Patrick, R. 1949. A proposed biological measure of stream conditions based on a survey of the Conestoga Basin, Lancaster County, Pennsylvania. Proc. Acad. Nat. Sci. Phila., Vol. 101, pp. 277-341.

——————. 1961. A study of the numbers and kinds of species found in rivers in eastern United States. Proc. Acad. Nat. Sci. Phila., Vol. 113, pp. 215-258.

——————, J. Cairns, Jr. and S. S. Roback. 1967. An ecosystematic study of the fauna and flora of the Savannah River. Proc. Acad. Nat. Sci. Phila., Vol. 118, No. 5, pp. 109-407.

Picken, L. E. R. 1937. The structure of some protozoan communities. Jour. Ecol., Vol. 25, pp. 368-384.

Robertson, T. B. 1921. The influence of mutual contiguity upon reproductive rate and the part played therein by the "x-factor" in bacterized infusions which stimulate the multiplication of infusoria. Biochem. Jour., Vol. 15, pp. 1240-1247.

——————. 1924. Allelocatalytic effect in cultures of Colpidium in hay-infusion and in synthetic media. Biochem. Jour., Vol. 18, pp. 612-619.

Rodhe, W. 1948. Environmental requirements of fresh-water plankton algae. Symbolae Bot. Upsaliensis, Vol. 10, pp. 1-149.

Schlichting, H. E., Jr. 1961. Viable species of algae and Protozoa in the atmosphere. Lloydia, Vol. 24, No. 2, pp. 81-88.

——————. 1964. Meteorological conditions affecting the dispersal of air-borne algae and Protozoa. Lloydia, Vol. 27, No. 1, pp. 64-78.

Spoon, D. M. and Burbanck, W. D. 1967. A new method for collecting sessile ciliates in plastic petri dishes with tight-fitting lids. Jour. Protozoology, Vol. 14, No. 4, pp. 735-744.

Stillwell, R. H. 1962. A stimulatory effect on the growth of *Paramecium caudatum* by products of *Colpidium campylum*. M.S. dissertation. St. John's University, Jamaica, New York.

—————————. 1967. Colpidium—produced RNA as a growth stimulant for Tetrahymena. Jour. Protozoology, Vol. 14, No. 1, pp. 19-22.

Zimmerman, P. 1961. Experimental untersuchungen uber die okologische Wirkung der Stromungsgeschwindigkeit auf die Lebensgemeinschaften des fliessenden wassers. Schweiz. Zeit. Hydrologie, Vol. 23, pp. 1-81.

J. K. G. Silvey and J. T. Wyatt

The Interrelationship Between Freshwater Bacteria, Algae, and Actinomycetes in Southwestern Reservoirs

Abstract

Some specifics and generalities regarding interrelationships of freshwater microflora are discussed. Particular emphasis is given the aquatic actinomycetes and the planktonic blue-green algae of southwestern reservoirs. The occurrence, distribution, enumeration, and some aspects of the overall physiology of aquatic bacteria, actinomycetes, and planktonic algae are reviewed. Also, suggestions are made relative to the role of each organism in such facets of reservoir metabolism as uptake, turnover, nutrient cycling, and organic production. Some speculative observations regarding competition and/or enhancement between these organisms are tentatively offered.

Introduction

Information on the interrelationships between different varieties of organisms composing the microflora of reservoirs is based largely on data secured from Lake Hefner in Oklahoma City and Garza Little Elm Reservoir near Denton, Texas, although additional reservoirs, streams and ponds are considered. Extensive studies have been made on Lake Hefner since 1952. During the summer period studies were carried out on a weekly basis and during the winter period on a monthly basis. Perhaps Lake Hefner has been studied in more detail than any reservoir in the southwest. As early as 1952 investigations were being made on the heat budget of that particular reservoir because of its unusual construction and location. Lake Hefner is an off-set reservoir located four and a half miles from the North Canadian River; it receives its water by way of a canal, which is controlled by gates during flows in the river. Since it is off-set and therefore not subjected to floods, the quantity of water added to the lake can be metered. Measurement of the amount that is lost either by way of evaporation or through a venturi as it enters the water plant makes this lake an unusual reservoir for study. Thus it has offered excellent opportunities for investigations on heat budget for evaporation control techniques, and for microbiotic cycles of a southwestern reservoir.

Garza Little Elm Reservoir, located approximately nine miles from Denton, Texas, obtains its water from three streams and exhibits a variety of ecological areas quite different from those found in Lake Hefner. These two reservoirs afforded an opportunity for a comparative study of the microbiotic cycles in an on-stream reservoir contrasted with an off-set type over a period of almost twenty years. Moreover there is diversity in depth between the two reservoirs. Most of Garza Little Elm is relatively shallow, roughly five to six meters, the deepest area being 15 meters. Lake Hefner, which was excavated into an almost circular basin, has an average depth of nine meters and a surface area of 2,500 acres. A comparison of the microorganisms and their cyclic phenomena in these two reservoirs should fairly well establish the ecological parameters obtained from intensive qualitative and quantitative limnological investigations heretofore neglected in this area of the nation.

As pointed out by Cole (1963), limnological information from the Texas area has not been very profound. Some qualitative information, but not very much quantitative data, has been accumulated from a study of its streams and reservoirs. It is the purpose of this paper to describe the observations and attempt to account for the microbiotic relationships in the warm water reservoirs as demonstrated by detailed studies of planktonic algae, freshwater bacteria, and aquatic actinomycetes. While some attention has been given to zooplankters, fungi, and benthic organisms, the data are not sufficiently detailed to be included. It is perhaps noteworthy that the concentration of protozoa, microcrustacea, and rotifers in southwestern reservoirs is very low when compared with data from lakes in Michigan, Minnesota and Wisconsin. According to our data, benthic organisms, on the other hand, are less sparse. In the shallow regions of our reservoirs, they comprise a population greater than is found in similar areas around Douglas Lake in Michigan or Lake Mendota in Wisconsin.

One of the first comprehensive studies done on reservoirs in the southwest was completed by Harris and Silvey (1940) concerning four reservoirs in the northeastern section of Texas. This was followed by a paper published by Cheatum, et al., (1942) on a single impoundment in the eastern part of the state. Deevey (1957) briefly examined some ponds in the Texas coastal plains and the arid Trans-Pecos Texas region. In the Oklahoma region, contributions have been made by Dorris (1956, 1958) concerning areas of the middle Mississippi and adjacent waters. Since that time, he has completed a number of studies in the state of Oklahoma. Irwin (1945) contributed information concerning the precipitation of colloidal particles from turbid waters in various areas of Oklahoma. There are numerous papers on fisheries from both states and in other areas of the Southwest but very little information of a quantitative nature having to do with the microbiotic cycles and the interrelationship of the organisms in reservoirs. Those publications that might be noted are largely concerned with the identification of species but with no indices of environmental conditions from which they came.

Discussion

The predominant types of algae occurring in the reservoirs of the southwest are greens, blue-gleens, and to a certain extent, flagellates and diatoms. To be certain, there are other varieties but they normally do not comprise a large proportion of the algal population. It is interesting to note that most of the green algae are unicellular or small colonial types. Larger planktonic forms do not normally occur in southwestern reservoirs. It is unfortunate that our data do not include the rather large quantities of sessile algae that are found growing along the perimeter of the reservoirs throughout all seasons of the year. We have not developed a technique for quantitation of these organisms, which perhaps comprise a greater biomass than the planktonic forms and perhaps contribute materially to the microbiotic cycles. We realize this deficiency but have not devised a method of surmounting the difficulty of determining the quantities of this material available in reservoirs. The cyclic phenomena of the various types of algae will be described in a later section. Details will be presented concerning the genera available, their respective concentrations, and the effects of chemical and environmental conditions upon the growth of these organisms.

The bacteria comprise a group of approximately ten common genera. Most of these bacteria are also found in soil and frequently in greater concentrations than they are found in the waters of reservoirs. However, the density of the population in reservoirs is much greater than in natural lakes in the middle west and northeast. From time to time, various types of pathogens are encountered. This is anticipated in reservoirs in the southwest since effluent from sewage disposal plants, run off from feed lots, and water-shed contamination frequently contribute these organisms, although they occur only on a sporadic basis. This situation is dependent upon the location of the reservoir and the source of its water. More details will be given concerning the cyclic phenomena of the bacteria and their relationship to the actinomycetes and algae.

The other group of organisms to be considered are the aquatic actinomycetes that have been isolated from reservoirs and studied both in laboratory and field investigations. Normally two genera are encountered, namely *Streptomyces* and *Micromonospora*. On rare occasions one may isolate a member of the genus *Nocardia*, especially from bottom deposits. The actinomycetes are an interesting group of organisms and have been studied in our laboratory since 1948. The above organisms comprise the population of microflora that will be considered in their interrelationships in the two types of reservoirs and other aquatic environments that have been studied in detail.

Planktonic Algae

Between population peaks, the planktonic algae of southwestern reservoirs are usually not particularly conspicuous. Discounting the usual slight variations, about fifty percent of the total population will be made up of green algae, and the remainder composed of sparse numbers of diatoms, occasional pyrrophytes, with the balance consisting of a small group of blue-greens. For the most part, the green algal population consists of unicells or non-filamentous forms belonging to the genera *Ankistrodesmus, Chlorococcum, Coelastrum, Oöcystis, Pediastrum, Scenedesmus*, and *Selenastrum*. Upon observation of most of a population in a particular counting field, notation of more than ten different green algal genera is unusual. Concerning the other major phytoplankton component, records indicate the average blue-green contingent to be fewer in number than the green algal population and generally composed of relatively inconspicuous forms. These are usually the small and somewhat difficult to identify organisms such as *Anacystis, Borzia, Gomphosphaeria, Phormidium*, and *Pseudanabaena*. We have not studied the other planktonic algal groups in sufficient detail to merit further elaboration.

Because thorough laboratory investigation of all planktonic algal groups seemed unfeasible, our studies have been almost exclusively limited to the blue-green algae. This restriction is perhaps unfortunate, but it has permitted detailed observation of most of those "bloom-formers," which at various times almost completely dominate the planktonic community. Also, it is quite well known that much reservation must be maintained when projecting data compiled in the laboratory back into the aquatic system. However, some evidence is beginning to accumulate that basic metabolic behavior patterns of phytoplankton may be much the same under laboratory conditions as *in situ*. For example, Watt (1966, 1969) reports little difference in release of extracellular products from forms grown naturally or laboratory-cultivated.

When compared to their terrestrial or amphibious counterparts, the planktonic blue-green algae tend to be highly active. Measurement shows such abilities as nitrogen-fixing (when applicable), photosynthetic, and respiratory to be much higher in euplanktonic blue-greens. This may not be due as much to genetically determined enzymic differences as to actual growth form. Non-planktonic blue-green algae seem to invariably grow either attached, or in mats, clumps, or balls. Since only a small portion of the algal mass is usually exposed to light and/or active substrate, physiological and biochemical measurements are generally low per amount of algal material. On the other hand, planktonic blue-greens generally remain as evenly suspended unicells, hormogones, or microscopic groups. The apparently much greater abundance of gas vacuoles may be a factor in maintaining this mode of life. Nonetheless, this greater activity of planktonic blue-greens is real under our laboratory conditions. From an ecological point of view, the reasons for this phenomena are probably not too important.

We have noted at least eight species of blue-green algae as major components of phytoplankton population peaks in area reservoirs. These include *Anabaenopsis circularis*, *Oscillatoria Agardhii*, *Nodularia Harveyana*, *Anabaena flos-aquae*, *Anabaena Bornetiana*, *Anabaena circinalis*, *Aphanizomenon flos-aquae*, and *Microcystis aeruginosa*. No strain of *Microcystis aeruginosa* yet isolated has ever been reported to fix atmospheric nitrogen. Also, the only culture of *Aphanizomenon flos-aquae* that we have been able to isolate and culture in the laboratory is unable to grow in nitrate-free medium or reduce acetylene (a usually reliable test of nitrogen-fixing ability). Obviously, some strains of *Aphanizomenon flos-aquae* are true nitrogen-fixers (Stewart, et al., 1968; Gentile and Maloney, 1969) while others are not (Williams and Burris, 1952). Since our particular *Aphanizomenon* is highly susceptible to even the slightest environmental stress, our data may be inconclusive. It might also be interesting to note that the 20 or so species or strains of planktonic blue-green "non-bloom-formers" which we have been able to isolate into unialgal culture are all apparently unable to fix atmospheric nitrogen. From this, one might suggest that the ability of an alga to fix atmospheric nitrogen is a prime factor in permitting it to become a dominant part of the phytoplankton. On the other hand, our strains of *Microcystis* and *Aphanizomenon* were both isolated from small "blooms" and neither can fix nitrogen.

Any of the aforementioned "bloom-forming" blue-green algal species might be almost the only, or most conspicuous, "waterbloom" component. However, our peak populations of blue-greens (which would probably never be afforded "bloom" status if compared to northern lakes) are usually mostly composed of three species. *Anabaena circinalis* generally seems to be most often the major component (about 50% of the total) with *Microcystis aeruginosa* and *Aphanizomenon flos-aquae* equally forming the balance of the population.

Although we seem to have made almost no progress in establishing absolute criteria for occurrence of "waterblooms" on southwestern reservoirs, it might be worthwhile to note the following. Observations compiled over the last several years indicate that most "waterblooms" seem to have developed after a severe or abrupt change in reservoir conditions. These changes have often been recorded in the form of rapid additions or losses of large volumes of water after major weather changes.

In laboratory culture, blue-green algal uptake of organic substrate (Allison et al., 1953; Carr and Pearce, 1966) does not seem to differ significantly from other types of algae (Sloan and Strickland, 1966). However, algal uptake of organic substrate may be relatively unimportant at the low concentrations found in the natural environment (Wright and Hobbie, 1966; Hobbie, 1966; Munro and Brock, 1968). This is in contrast to the substantial amounts of extracellular materials produced by algae (Fogg and Westlake, 1955; Fogg et al., 1965; Hellebust, 1965, 1967; Watt, 1966, 1969).

We have not been able to show a significant increase in planktonic blue-green algal growth, either heterotrophically or autotrophically with a wide variety of organic substrates (acetate, citrate, inositol, glucose, sucrose, and casamino acids). These have been tested at concentrations up to .01 molar on both axenic and non-axenic unialgal cultures. These negative results seem to confirm the fact that algal uptake of organic subtrate, particularly at the normally low concentrations found in our reservoirs, is not significant. For example, preliminary tests during the spring of 1969 or dissolved glucose concentrations in Garza Little Elm Reservoir only indicated a concentration on the order of 10^{-10} molar.

Bacterial enumeration on area reservoirs usually reveals a rather abrupt population increase following algal population peaks (Fig. 1). Whether or not this is presumptive evidence that extracellular organic production is a significant factor in the heterotrophic plankton population has not been established. Laboratory studies (which are discussed in a following section) do not indicate pronounced support of increased bacterial populations by extracellular algal production. Since no accumulation of products is permitted in the culture vessel, this observation may not be pertinent to actual reservoir conditions.

The relatively sudden disappearance of a major component of an algal population from the reservoir has not always been adequately explained. It would seem that perhaps many interrelated factors must be involved. Observations on laboratory cultures does not seem to offer an easy solution either. Sometimes seemingly healthy non-axenic cultures of a generally hardy blue-green alga will die almost overnight. Instead of the characteristic bright orange-yellow color of nutrient starved blue-green algae (Kingsbury, 1955; Venkataraman, 1960), the cultures will rapidly

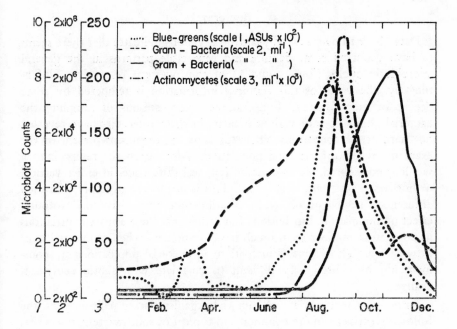

FIGURE 1. Typical microbiotic cycle of a southwestern reservoir.

become almost colorless with a marked whitish turbidity, which is probably due to rapid bacterial growth and indicates disintegration of the algal cells.

On the other hand, selective removal of dietary components from a blue-green algal laboratory culture will cause quick deactivation of the culture, but not death. Within a few days, the blue-green culture will chloress, turn orange-yellow, and finally to a dirty brown. These minerally deprived organisms along with their associated flora have been maintained under normal laboratory culture conditions for periods up to one year. When the missing nutritive element was added complete recovery of the entire cell mass occurred within three days.

Would these organisms have been decomposed in the reservoir during this time? Since actinomycetes were not present, and the surviving bacterial flora might not have been active decomposers, this observation may not carry over into the reservoir. However, it serves to indicate that algal decomposition products may not be as easily and as quickly available to the heterotrophic population as is usually implied.

Bacteria

Data obtained over several years' observations indicates that there seems to have occurred some definite and repeatable patterns in the general microbiotic cycles (Figs. 3 and 1) of southwestern reservoirs. Unfortunately, the value of this acquired information is tempered by some degree of inconsistency. For example, many seasons of sampling the bacterial populations of various habitats in these reservoirs have revealed no "hard" rules concerning either the types of organisms present or the exact magnitude of the total population. Also, when the reservoir as a whole is considered, we have noted few real differences in either vertical or horizontal distribution (Fig. 2). This is not meant to imply that such differences never occur (e.g., stratification may have a profound effect on population distribution) but rather that these apparent variations in population number seem much less pronounced when monitored over extended periods of time. Fluctuations in bacterial populations are common, but have been very difficult to correlate with subtle ecosystem changes.

Although a wide variety of bacterial organisms are usually present, the following types are more common and typical of southwestern reservoirs.

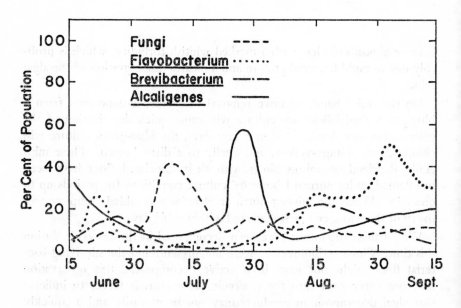

FIGURE 2. Dominant heterotrophs in Lake Hefner (summer 1967).

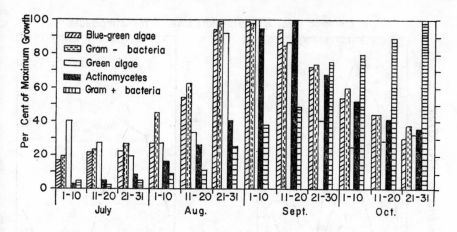

FIGURE 3. Relative summer population of a typical southwestern reservoir.

The gram-positive variety include *Bacillus cereus, B. cereus* var. *mycoides, Brevibacteria,* and less frequently, the streptococci (Group D or entero-cocci) which are usually recovered most frequently at "influx" stations. The gram-negative organisms usually make up the greater portion of the bacterial flora of reservoirs for most of the year (Figs. 2 and 1). The dominant forms in this group are normally the genera *Flavobacterium, Pseudomonas, Alcaligenes,* and some members of the *Enterobacter-Klebsiella* group, which, again in their occurrence, usually reflect "influx" stations.

Many articles have discussed the advantages and disadvantages of different types of bacterial enumeration. In this country plate counts have generally been favored whereas in Europe the direct count method has been used most frequently. Since about 1930 (Potaenko, 1968), Russian workers have employed direct count methods in most investigations. As Fig. 4 indicates, these methods demonstrate a much higher population of organisms than standard plate count methods. Rodina (1967) stated that only 10^{-4} of the entire population is shown by plate counts in unpolluted waters but 10^{-2} is revealed in polluted waters. However, Straskrabova and Legner (1969) had about equal success using both direct and plate counts in estimating bacterial numbers in running and stagnant waters and correlating these to five-day biological oxygen demand measurements. Bere (1933) used direct counts to survey the bacterial population of Wisconsin lakes. Later, Henrici (1939) considered Bere's high counts (up to 2×10^6/ml), obtained by the direct method, unreliable. Currently, con-

1. eutrophic lake *2.* intermediate lake *3.* mesotropic lake

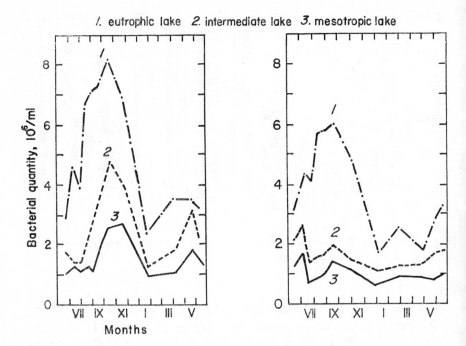

FIGURE 4. Comparative bacterial population via direct count after Potaenko, 1968.

siderable attention is being devoted to membrane filter-fluorescent-antibody techniques (Guthrie and Reeder, 1969) and luminescent microscopy (Rodina, 1967).

Since our reservoirs contain substantial quantities of particulate detritus which severely hinders accuracy in direct counting, we generally use the spread-plate technique. Its convenience and avoidance of thermal shock, since many indigenous aquatic microbes may be quite stressed by the 45 to 50 C. temperatures (Wierings, 1968) of molten agar used with the pour plate method, has made it more attractive than the standard pour plate method.

When microbial population counts of southwestern reservoirs (Fig. 5) made with spread plates are compared to direct counts (Fig. 4), the former are usually low and exhibit much the same magnitude as the values obtained from plate counts by Henrici (1939) and Potter (1964). It is apparent that a clearer picture of the nature of the bacterial population is obtained by incubating at both 37 and 24 C. and maintaining differential counts to five days. As Fig. 6 clearly demonstrates, if plates were read at the end of a standard 48-hour period, neither of the incubation temperatures would yield a valid picture of the actual bacterial population.

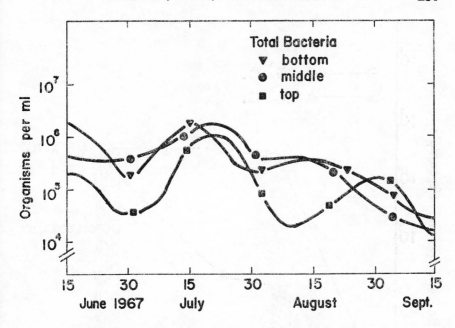

FIGURE 5. Vertical distribution of bacteria (Lake Hefner).

Even after five days, our ratio of chromogens (which are usually considered as indigenous flora) does not exceed 50% and is somewhat lower than values obtained by Potter (1964).

The exact extent of bacterial action in the reservoir environment is not easily quantitated. Generally, the process of organic nutrient substance cycling has been indirectly estimated by correlating specific bacterial and actinomycetic counts with reservoir conditions with which they seemed to be associated. It has been only in the last few years that we have attempted measurements of specific nutritive components and/or processes *in situ*. Some of these deserve specific description. For example, by use of C^{14} carbonate, we have initiated comparative studies of autotrophic and heterotrophic uptake and extracellular release in several southwestern impoundments. Attempts are being made to correlate many of these relatively recent measurements with the large mass of traditional limnological data gathered over the last two decades. Final computation of data is incomplete at this writing. The recent development (Stewart, et al., 1967, 1968) of the acetylene-reduction technique has made indirect measurement of nitrogen-fixation quite practical. So far, we have not been able to detect any free-living aquatic-heterotrophic bacterial reduction

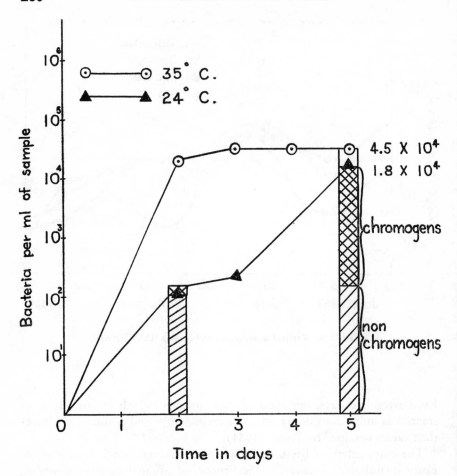

FIGURE 6. Bacterial enumeration by the spread plate technique with incubation temperature as a counting variable.

of acetylene. Thus, this preliminary evidence suggests that, at least in southwestern reservoirs, only blue-green algae are important in this regard.

Organic productivity measurements on area reservoirs are made by the carbon dioxide (Beyers et al., 1963) and oxygen (Odum and Hoskins, 1958) rate-of-change methods. Generally our southwestern reservoirs cannot be considered high in productivity, which is perhaps limited by excessive turbidity. An extensive study of Garza Little Elm Reservoir produced an average net production of 60 g $CO_2 \times m^{-2} \times day^{-1}$ fixed (Trotter, 1969).

It is now well documented that a considerable portion of the total algal production is released as extracellular products (Fogg, 1966), both in culture and natural environments. We have only been able to indirectly follow the effect or utilization of these extracellular products in laboratory cultures of blue-green algae and their naturally associated bacterial flora. To accomplish this, three nitrogen-fixing planktonic species (*Anabaena Bornetiana, Anabaena circinalis,* and *Anabaena flos-aquae*) were isolated into unialgal culture by serial dilution in combined-nitrogen-free inorganic medium from natural population peaks in area reservoirs. Applying no other purification procedures, we have maintained these cultures in log phase by transfer of 1 ml inocula each time the O.D. 650 reached about .006 (generally after about five days growth). Although original bacterial counts were high (about the same as the natural water from which the original isolate was made), after three or four transfers, the bacterial count was negligible. Subsequent tests for antibacterial exudates in the culture medium proved negative. The bacterial populations increased only when the algal culture was permitted to age or mature. We tentatively suggest that, at least for our laboratory conditions, normal extracellular algal production is not of sufficient types or concentrations to promote extensive bacterial development. Also heterotrophic bacteria may be more dependent on allochthonous organics and decomposition products of plankton (Fig. 7) than extracellular products. Whether or not this is true in reservoirs remains to be shown.

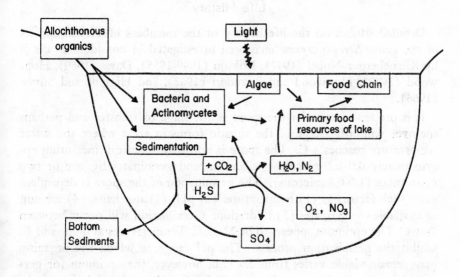

FIGURE 7. Primary production and microbiological processes in a deep lake after Sorokin (1968).

Aquatic Actinomycetes

One of the early references to aquatic forms of the actinomycetes was made by Adams (1929). Burger and Thomas (1934) studied the tastes and odors in the Delaware River and while not denying the involvement of actinomycetes, indicated that the odors resulted from the activity of a mixture of microorganisms. Thaysen (1936) reported salmon of the Thames River acquired an earthy or muddy taint as a result of actinomycete-produced compounds from aquatic types. Isachenko and Egorova (1944) published the results of earlier work on several rivers in Russia where earthy odors were attributed to the actinomycetes predominantly growing, they thought, on the bottom and sides of the rivers. Silvey, et al., (1950), showed that certain actinomycetes were responsible for a variety of tastes and odors by comparing the odor produced from pure cultures of organisms with the odors found in water supplies. Later laboratory investigation by Dill (1951), McCormick (1954), and Pipes (1955) revealed that certain types of compounds produced by these aquatic actinomycetes might have some relation to the microbiotic cycle as it occurred in the reservoirs, streams and ponds that they investigated. In the last six years several papers have appeared which describe methods of growing various species of the genus *Streptomyces*, extracting odor components, and studying them for chemical identity.

Life History

Detailed studies on the life history of the members of certain species of the genus *Streptomyces* have been investigated in considerable detail by Klieneberger-Nobel (1947), Erikson (1949-1955), Davis (1960), Hopwood (1960), Hopwood and Sermonti (1962), and Higgins and Silvey (1966).

It is proper to consider the spore as the stage of transfer and perhaps the over wintering stage of the aquatic forms in areas where the water temperature reaches 4 C. The spore is normally spherical, measuring approximately 0.9-1.2 microns in diameter, and germinates by one or two germ tubes (1.0-0.8 microns). The germination of the spore is dependent upon such factors as (1) temperature, (2) pH, (3) nutrients, (4) amount of available oxygen, and (5) hydration. Germination will occur between 7-36 C. The optimum appears to be 22-35 C. Temperatures as high as 65 C. inhibit the germination process. The pH range at which the organism may remain viable varies from 2.0 12.0; however, the optimum for germination appears to occur between 6.5-8.5. When the richness of the medium was increased, there was usually an increase in the time and level of total primary mycelia production. Germination times appeared to be very little affected by the varying nutrient strength. It should be noted that germination and development were inhibited when the total solids exceeded 60 grams per liter. Observations made by Higgins and Silvey (1966) showed that the organism favored an aerobic pathway, since anaerobic conditions greatly reduced the rate of growth. It was noted in laboratory as well as in field work that the germination and development of the primary mycelium required high levels of hydration. Dessication was found to be the most successful method of controlling the amount of primary growth in all laboratory investigations (Higgins and Silvey, 1966).

After spore germination and before branching began in the primary mycelium, there was a lag period. Then the branches started as swellings at the sides of a hyphae. These swellings extended laterally while some maintained a sort of a bulboid appearance. Septation was not common in the primary mycelium, especially in the early growth stages. As the growth of the primary mycelium continued, the diameter of the newly formed hyphae ordinarily diminished to about 0.4-0.6 microns. It is interesting to note that the branching was not limited to the plane of the medium, but rather many of the filaments penetrated deeply into the substrate while others became flattened onto the medium so as to maintain sufficient hydration.

The primary mycelium frequently propagated itself vegetatively by fragmentation. Reproduction of this nature may not be commonly observed due to the fact that the primary mycelium is unusually small. If, however, in the laboratory experimentation, primary mycelia undergo fragmentation by extreme agitation many colonies appear in the liquid medium, indicating an origin of new primary mycelia by this system.

The secondary mycelium differs from the primary in that it is larger, usually carries a black, insoluble pigment in the outermost cellular layer, that it is less branched and will not attempt to penetrate a solid medium. In addition, one may note that if the secondary mycelium undergoes fragmentation small fragments of the hyphae revert to a primary stage. If several hyphae are involved, new secondary mycelia are produced.

Spores are produced at the apices of the hyphae. All notations on the secondary mycelium indicate that in its final development prior to and during sporulation, it becomes somewhat hydrophobic in liquid culture and floats to the surface. It is interesting to note that as the secondary mycelium developed, the primary mycelium which seems to serve as part of the nutrient for the growth of the secondary ones, undergo slow autolysis.

In reservoirs the primary stages may occur near the perimeter of the lake and may be associated with emergent, submerged, or floating vegetation. Also, since they are facultative aerobes, they may grow on the bottom mud during periods of summer stratification. As long as temperatures are 20 C. or below, the primary stage apparently persists in that condition, although in instances where water is drawn down in a reservoir so as to expose the primary stages near the perimeter of the lake, as the hydration decreases, secondary mycelia are formed. So long as the primary mycelia remain submerged, and the temperature remains at about 20 C, they do not appear to develop, and thus secondary mycelial formation seems unlikely. As the temperatures increase in the water, the primary stages that are in the aerobic zone of the reservoir and are associated with nutrient sources such as higher aquatic plants, blue-green algal mats, or organic ooze, produce secondary mycelia around 22 C. As the water becomes warmer, more secondary stages form. During early fall turnover, as the hypolimnion of the lake becomes aerobic, primary mycelia are altered to secondary mycelia. These come to the surface of the water along with mats of floating blue-green algae, and are slowly carried shoreward. In reservoirs with dense areas of emergent vegetation, or the remains of submerged trees (quite abundant in the southwest), the formation of secondary mycelia may be anticipated during warm periods beginning about the middle of July, and may well continue for some time depending upon meteorologic conditions (Silvey, 1964).

Cultivation and Enumeration

The isolation and enumeration of aquatic actinomycetes may pose numerous problems due to the fact that spores and hyphal elements of both the primary and the secondary mycelia will all develop colonies on isolation plates. It should therefore be understood that the counts of the actinomycete population made from reservoirs or streams reflect a relative density but not a true number as might be found perhaps in bacterial isolation, cultivation, and enumeration techniques.

Standard bacteriological techniques are involved in the collection of samples from reservoirs, ponds, or streams, and for isolation spread plates are used rather than poured plates. This is due to the fact that more rapid appearance will be noted in the secondary mycelial colonies. The period of incubation involves from six to fourteen days. Since there are numerous bacteria mixed in with the actinomycetes, it is desirable to incubate part of the plates at room temperature and others of the same dilution at 32° C. Thirty-seven degrees centigrade frequently causes the bacteria to overgrow the plates and dehydration of the medium will occur prior to the time the actinomycetes appear.

A number of types of media that may be employed for the isolation of actinomycetes from water samples are available. One on the market as a commercial product is Actinomycete Isolation Agar, which appears to be quite successful. Another variety that we have employed from time to time is Emerson's modified agar which contains 20 grams of agar, 80 grams brown sugar, 8 grams peptone, 4 grams beef extract, 2 grams meat extract, 2 grams sodium chloride made to 1,000 milliliters in water. Regular commercial plate count agar, when enriched with 20 grams brown sugar per 1,000 milliliters, makes an excellent isolation agar for the aquatic streptomyces. If it is desirable to have a minimal medium which may be used in instances where slow growth is desirable and which will not be highly contaminated by bacteria, a formulation comprised of the following ingredients should be employed: sodium citrate, 10 grams; sodium nitrate, 2 grams; potassium nitrate, 2 grams; calcium chloride, 0.1 gram; magnesium sulfate, 0.05 grams; bipotassium phosphate, 2 grams; agar, 20 grams; distilled water to make 1,000 milliliters.

Nutrient Sources for Actinomycete Development

It is well recognized that the actinomycetes have a rate of growth normally slower than most of the fungi and that they do not reproduce as rapidly as most of the bacteria. However, this group of organisms does have the attribute of producing antibiotics and these substances may enable them to compete successfully with more vigorously growing types of aquatic microorganisms. Members of the genus *Streptomyces*, after once colonizing on some type of substrate, are very persistent in their growth for a considerable period of time. In addition, these organisms share with the fungi in the ability to grow into substrates. Although it is recognized their hyphal elements are highly delicate as compared to the fungi, they can obtain sufficient penetration for continuous nutrition.

The role of the actinomycetes in biodeterioration has been well summarized by Williams (1966). A perusal of his paper leads one to conclude that members of the genus *Streptomyces* may make use of simple sugars as well as polysaccharides. The ability of the members of this genus to produce cellulose makes them important organisms in the microbiotic cycle in the reservoirs where cellulose deposits lead to rapid eutrophication. Moreover, the ability to deteriorate lignin is also important in the reduction of the rate of eutrophication in reservoirs. It is further noted by Williams (1966) that certain lipids and steroids may serve as carbon sources and they may be either partially or completely degraded. In addition, tannins of low molecular weight, including tannic acid and gallotannin, may be decomposed by soil-inhabiting actinomycetes.

Inorganic sources of nutrition include nitrogen containing compounds such as ammonia and nitrates. In the previously listed media it should be noted that either ammonia or nitrates are added in order to encourage actinomycetic growth. That actinomycetes apparently need ample amounts of combined nitrogen is demonstrated by their repeated occurrence as contaminants only in blue-green algal isolates which are grown in complete medium. We have not yet detected actinomycetic growth in cultures of blue-green nitrogen-fixers which are routinely grown in nitrate-free medium. Other nitrogen sources may include organic types, since proteolytic activity of the actinomycetes appears to have been well established, as pointed out by Williams (1966). Organisms that contain such a vast array of enzyme systems and metabolic pathways are obviously very important in numerous phases of a microbiotic cycle that occurs in freshwater reservoirs.

Composition of Streptomyces Isolated from Reservoirs

The different species of the genus *Streptomyces* isolated from the area reservoirs have been identified by serological and electron microscopic examinations (Tables 1 & 2). No attempt has been made to do detailed morphological and nutritive studies in order to accomplish identification.

TABLE 1.—Results of Serological and Electron Microscopic Examination of Twenty Unknown Species of Streptomyces.

Culture Number	Spore Surface	Species Identification	Serological Similarity*	ISP Reported Spore Surface†
H1	smooth	*antibioticus*	4	smooth
H9, H20, H21	smooth	*odorifer*	4	smooth
H24, H34	spiny	*viridochromogenes*	4	spiny
H10, H13, H32	smooth	*aureofaciens*	4	smooth
H8	warty	*aureofaciens*	3	smooth
H17	spiny	*aureofaciens*	1	smooth
H2, H3, H21	warty	*griseolus*	2	smooth
H11	smooth	*coelicolor*	2	smooth
H28	spiny	*coelicolor*	1	smooth
H7	smooth	*coelicolor*	1	smooth
H27, H29	spiny	*odorifer*	3	smooth
H30	warty	*odorifer*	1	smooth

*Serological relationships, 1 = related, 2 = close, 3 = very close, 4 = identical
†Shirling and Gottlieb, 1968; from Taylor and Guthrie, 1968

TABLE 2.—Recent Isolates of Odor-Producing Streptomyces at NTSU.

Culture Number	Species Identification	Serological* Similarity
SH9	Streptomyces cinnamoneus	1
Rat	Streptomyces coelicolor	4
62	Streptomyces antibioticus	4
NT16	Streptomyces antibioticus	4

*Serological relationships: 1 = related; 2 = close; 3 = very close; 4 = identical. <

In all instances where serological techniques were involved known identified species were used against the unknown in order to obtain information concerning the serological similarity. The species involved that were identical in all serological tests are *Streptomyces antibiotics, S. odorifer, S. virivochromogenes, S. aureofaciens, S. cinnamoneus,* and *S. coelicolor.* References made in this paper to *Streptomyces* may be to any of the above mentioned species since the cultural differences were not great except in the spore form, shape, pigmentation, and in the color of the outer cellular layer. No detailed data on any one species can be quoted at this time.

Cyclic Phenomena in Southwestern Reservoirs

The actinomycete population density has been measured for a number of years in Lake Hefner, Garza Little Elm, water supply reservoirs for Wichita Falls, Graham, Abilene and other West Texas towns. On occasion, in order to obtain information concerning the population density of actinomycetes, samples have also been analyzed from streams that supply the reservoirs. These samples have been collected during times of both high and low water. The population density of these samples has been recorded as a mean and noted on a monthly basis. It may be observed in Table 3 that the actinomycete population usually begins to increase in the top and the bottom of reservoirs during the month of July, reaching a peak in August. Obviously the high counts obtained from bottom deposits result mostly from spores. Also, many of these may be of recent origin due to the diversity in the population density as demonstrated by the colony count. These cycles demonstrate rather clearly a period of increased growth and a period of diminution. During September most reservoirs in the southwest "turn-over". This gives more equalization in the distribution of the actinomycetes in the reservoirs. In November the population begins to decrease, so from December through spring, the concentration of organisms is relatively low.

TABLE 3.—Actinomycete Population Density Average Southwestern Streams and Reservoirs.

Organisms/ml

Month	Jan.	Feb.	Mar.	Apr.	May	June
Avg. temp.	7.2°	5.6°	14°	16°	19°	22°
Top	100	80	90	120	340	760
Middle	40	20	40	38	46	60
Bottom	180	160	200	260	460	840

Month	July	Aug.	Sept.	Oct.	Nov.	Dec.
Avg. temp.	26°	29°	27°	24°	18°	10°
Top	1,160	120,000	30,000	12,000	800	190
Middle	85	400	35,000	20,000	1,200	200
Bottom	1,250	680,000	90,000	23,000	3,400	900

There is an obvious periodicity in the algal and actinomycetic blooms that occur in southwestern reservoirs. This may be modified from time to time by unusual flows during the month of June. Over a 15-year average, the June bloom appears to be relatively unimportant. If rainfall occurs in February, it is likely there will be a sporadic increase in the blue-green algae, followed by a growth of diatoms. This is again a rather seasonal phenomenon, but on an annual basis shown in Fig. 8 the tendencies are duplicated year after year. It is observed that blue-green algal peaks generally precede the "bloom" of actinomycetes, which is sometimes followed by an increased diatom population. The interrelationships

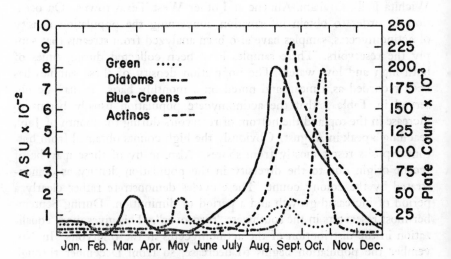

FIGURE 8. Cyclic phenomena in southwestern reservoirs.

between the blue-green algae and the actinomycetes are well demonstrated during this period and are duplicated year after year. Even in laboratory culture of blue-greens from these reservoirs it is quite common for secondary mycelia of the actinomycetes to grow and float to the surface of the culture chamber. This further indicates the close association that exists between these two groups of organisms, for in many cases it has proved almost impossible to rid the contaminated blue-green of its actinomycete "associate."

The bacteria also associated with the blue-green algae and the actinomycete blooms in southwestern reservoirs (Figs. 3 and 1) illustrate the average annual cycle of gram-negative heterotrophic bacteria which reach their maximum density at approximately the same time as the blue-green algae. Their association, of course, particularly in artificial culture, is very well recognized (Gorham, 1964; Vance, 1965). Following the blue-green algal bloom, the actinomycetes reach their maximum population density and in due time the gram-positive heterotrophic organisms undergo a rapid bloom which is attenuated normally by the last of December or early in January. These associations and cyclic phenomena have been observed over sufficient periods of time to be recognized as a possible interrelationship that bears further scrutiny. For example, in Fig. 9 the relationship between the gram-positive bacteria and the actinomycetes is very definite. As the actinomycete bloom begins to decrease the gram-positive population increases in a logarithmic curve from the early part of July to the middle of September. Work in the laboratory has demonstrated that the gram-positive bacteria, *Bacillus cereus*, is capable of metabolizing the odor components produced by the actinomycetes. We have not been able to show definitely whether they actually attack the hyphal elements of the streptomyces, or metabolize the organic components produced by these organisms. Detailed microscopic examinations reveal that the mycelial mass appears to be undergoing disintegration in proportion to the increase in population of *B. cereus*. After the actinomycetes have reached their peak of logarithmic growth, if there is an absence of these bacteria in laboratory culture, autolysis of the mycelia is very slow indeed. Since the organic components produced by the actinomycetes are recognized by their odors and the fact that they disappear from the water in the presence of high population concentrations of *Bacillus cereus*, it is obvious from our work that there is a metabolic and enzymic relationship between these two groups of organisms. We are hesistant to state precisely where or how it occurs.

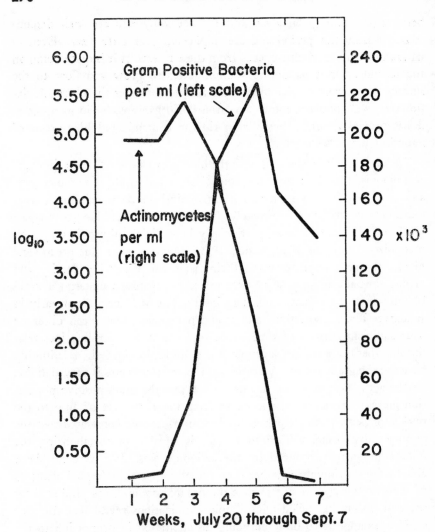

FIGURE 9. Inverse relationship between actinomycetes and gram positive bacteria.

Summary

The reservoirs of the southwest offer a variety of ecosystems, due to diversity in construction and maintenance. Since many areas are confined largely to agricultural practice while others are more or less industrialized, many varieties of nutrient materials enter the feed streams to the reservoirs. In general, one must consider these bodies of water as having relatively high quantities of nutrients and consequently should be unusually fertile. Added to these characteristics, the growing seasons are long and warm, encouraging certain species of the microflora to increase sporadi-cally. There is not normally a period of stagnation during the relatively mild winters. Therefore, the reservoir is in a "turnover" state from the middle of September until the middle of June. Perhaps the greatest factor limiting productivity in southwestern reservoirs is turbidity, which is normally attributable to various types of suspended inorganic matter, although from time to time organic detritus from the remains of sub-merged vegetation or inundated trees will produce both color and tur-bidity. Notwithstanding all of these various characteristics, one may consider southwestern reservoirs to contain a rather rich and varied plank-tonic microflora.

The phytoplankton are comprised of greens, blue-greens, diatoms, and occasionally, certain types of flagellates. The green algae appear to have an almost constant level of population density, with some variations in the spring and early fall. Diatoms may express a sudden increase or "water bloom" on occasions following high flows in the early part of the spring. Otherwise, their population is somewhat constant, rather diverse, and does not compose a large portion of the total population. In general, the phytoplankton picture in the southwestern reservoir reveals that blue-green algae exert the major influence upon the entire aquatic ecosystem, particularly during peak populations. These blue-green algae that exert such an important effect are dominated by nitrogen-fixing species. The euplanktonic blue-green algae have accelerated metabolic patterns when compared to other blue-green forms. Although these organisms do not seem to be highly involved in uptake of the organic substrate, they may produce substantial amounts of extracellular products. However, studies on young cultures of selected blue-green algae and associated bacteria indicate that the extracellular production by the algal population may be of limited use to the associated gram-negative heterotrophic flora. In laboratory culture, minerally starved blue-green algal organisms deactivate readily, but fail to die and decompose within a period up to a year. The gram-positive heterotrophic bacteria do not exhibit an increase in popu-lation density until the termination of an actinomycete bloom. Data secured from southwestern reservoirs show that the aquatic actinomycetes follow the blue-green algae bloom and laboratory observations indicate a more rapid deterioration of blue-green algae that have reached the logarithmic peak of growth in the presence of actinomycetes than in the absence of these forms; it could, therefore, be postulated that the bacteria in our aquatic environments are not contributing as much to the decom-position of the blue-green algae as the streptomyces which, in some man-ner, influence the reduction and perhaps degradation of the blue-green algae. In turn, from our findings, it appears that the gram-positive heterotrophic bacteria such as *Bacillus cereus*, which invariably increases after an actinomycetic bloom, may contribute to either the biogradation of the mycelial mat or to the by-products produced by the actinomycetes.

In addition, it must be recognized that the allochthonous organics and particulate detritus of an oxidizable nature that is brought into the reservoirs by way of the feed streams must be important in furnishing nutrients for the gram-negative and gram-positive heterotrophic bacteria. When the interrelationships of the microflora of the reservoir are better annotated, the problem of eutrophication may be more rapidly solved.

Acknowledgments

The investigations were supported in part by Federal Water Pollution Control Administration Grants WP-00785, WP-00805, and 5T1-WP-107; Bureau of Reclamation Contract 14-06-D-5298; the City of Oklahoma City, and North Texas State University water research laboratories.

Literature Cited

Abdirov, Ch. A., L. G. Konstantinova and N. S. Sagidullaev. 1968. Microbiology of Lake Karateren. Microbiology. 37:297-300.

Adams, B. A. 1929. Odors in the waters of the Nile River. Water and Water Engr. 31:309-314.

Allison, R. K., H. E. Skipper, M. R. Reid, W. A. Short and G. L. Hogan. 1953. Studies on the photosynthetic reaction. I. The assimilation of acetate by *Nostoc muscorum*. J. Biol. Chem. 204:197-205.

Bere, R. 1933. Numbers of bacteria in inland lakes of Wisconsin as shown by the direct microscopic method. Int. Rev. Ges. Hydrobiol. Hydrograph. 29:248-263.

Beyers, R. J., J. L. Larimer, H. T. Odum, R. B. Parker, and N. E. Armstrong. 1963. Directions for the determination of changes in carbon dioxide from changes in pH. Pub. Inst. Mar. Sci. Univ. Tex. 9:454-489.

Burger, J. W. and S. Thomas. 1934. Tastes and odors in the Delaware River. J. Amer. Water Works Assoc. 26:120-127.

Carr, N. G. and J. Pearce. 1966. Photoheterotrophism in blue-green algae. Biochem. J. 99:28p.

Cheatum, E. P., M. Longnecker and A. Metler. 1942. Limnological observations on an East Texas lake. Trans. Amer. Microscop. Soc. 61: 336-348.

Cole, G. A. 1963. The American southwest and middle America, p. 393-434. *In* D. G. Frey (ed.), Limnology in North America, U. of Wisconsin Press, Madison.

Davis, G. H. G. 1960. The interpretation of certain morphological appearances in a *Streptomyces* sp. J. Gen. Microbiol. 22:740-743.

Deevey, E. S., Jr. 1957. Limnological studies in middle America with a chapter on Aztec limnology. Trans. Connecticut Acad. Arts Sci. 39:213-328.

Dill, W. S. 1951. The chemical compounds produced by actinomycetes and their relation to taste and odors in a water supply. M.S. thesis, North Texas State University, Denton, Texas.

Dorris, T. C. 1956. Limnology of the middle Mississippi River and adjacent waters. II. Observations on the life histories of some aquatic Diptera. Trans. Illinois Acad. Sci. 48:27-33.

——————. 1958. Limnology of the middle Mississippi River and adjacent waters. Lakes on the leveed floodplain. Amer. Midland Nat. 59:82-110.

Erikson, D. 1941. Studies on some lake-mud strains of Micromonospora. J. Bacteriol. 41:277-300.

Fogg, G. E. 1966. The extracellular products of algae. Oceanogr. Mar. Biol. Ann. Rev. 4:195-212.

—————— and D. F. Westlake. 1955. The importance of extracellular products of algae in freshwater. Verh. Int. Verin. Limnol. 12:219-232.

——————, C. Nalewajko and W. D. Watt. 1965. Extracellular products of phytoplankton photosynthesis. Proc. Roy. Soc., B. 162:517-534.

Gentile, J. H. and T. E. Maloney. 1969. Toxicity and environmental requirements of a strain of *Aphanizomenon flos-aquae* (L.) Ralfs. Can. J. Microbiol. 15:165-173.

Gorham, P. R. 1964. Toxic algae, p. 307-336. *In* D. F. Jackson (ed.), Algae and man. Plenum Press, New York.

Guthrie, R. K. and D. J. Reeder. 1969. Membrane filter-fluorescent-antibody method for detection and enumeration of bacteria in water. Appl. Microbiol. 17:399-401.

Harris, B. B. and J. K. G. Silvey. 1940. Limnological investigation on Texas reservoir lakes. Ecol. Monogr. 10:111-143.

Hellebust, J A. 1965. Excretion of some organic compounds by marine phytoplankton. Limnol. Oceanogr. 10:192-206.

Hellebust, J. A. 1967. Excretion of organic compounds by cultured and natural populations of marine phytoplankton. Amer. Assoc. Advan. Sci. Publ. 83:361-366.

Henrici, A. T. 1939. The distribution of bacteria in lakes, p. 39-64. *In* F. R. Moulton (ed.), Problems of lake biology, Amer. Assoc. Adv. Sci. Publ. 10, Science Press, Lancaster, Penn.

Higgins, M. L. and J. K. G. Silvey. 1966. Slide culture observations of two freshwater actinomycetes. Trans. Amer. Microsc. Soc. 85:390-398.

Hobbie, J. E. 1966. Glucose and acetate in freshwater: Concentrations and turnover rates, p. 245-251. *In* H. L. Golterman and R. S. Clymo (eds.), Chemical environment in the aquatic habitat. Proceedings of an I.P.B.-Symposium in Amsterdam and Nieuwersluis 10-16 Oct. 1966. N. V. Noord-Hollandsche Uitgevers Maatschappij-Amsterdam.

Hopwood, D. A. 1960. Phase-contrast observations on *Streptomyces coelicolor*. J. Gen. Microbiol. 22:295-302.

Hopwood, D. A. and G. Sermonti. 1962. The genetics of *Streptomyces coelicolor*. Advan. Genet. 11:273-342.

Irwin, W. H. 1945. Methods of precipitating colloidal soil particles from impounded waters of central Oklahoma. Bull. Oklahoma Agr. and Mech. Coll., 42(11), 16 p.

Isachenko, B. L. and A. Egorova. 1944. Actinomycetes in reservoirs as one of the causes responsible for the earthy smell of their waters. Microbiologiya. 13:224-230.

Ivatin, A. V. 1968. Microbiological characteristics of the Kuibyshev reservoir in 1965. Microbiology 37:301-306.

Kingsbury, J. M. 1955. On pigment changes and growth in the blue-green alga, *Plectonema nostocorum* Bornet ex Gomont. Biol. Bull. 110:310-319.

Klieneberger-Nobel, E. 1947. The life cycle of sporing Actinomycetes as revealed by a study of their structure and septation. J. Gen. Microbiol. 1:22-32.

McCormick, W. C. 1954. The cultural, physiological, morphological and chemical characteristics of an actinomycete from Lake Waco, Texas. M.S. thesis, North Texas State University, Denton, Texas.

Munro, A. L. S. and T. D. Brock. 1968. Distinction between bacterial and algal utilization of soluble substances in the sea. J. Gen. Microbiol. 51:35-42.

Odum, H. T. and C. M. Hoskins. 1958. Comparative studies on the metabolism of marine water. Pub. Inst. Mar. Sci. Univ. Tex. 5:16-46.

Pipes, W. O. 1955. An investigation of naturally occurring tastes and odors from fresh waters. M. S. thesis, North Texas State University, Denton, Texas.

Potaenko, Yu. S. 1968. Seasonal dynamics of total bacterial number and biomass in water of Narochan lakes. Microbiology 37:441-446.

Potter, L. F. 1964. Planktonic and benthic bacteria of lakes and ponds, p. 148-166. *In* H. Heukelekian and N. C. Dondero (eds.), Principles and applications in aquatic microbiology, John Wiley, New York.

Rodina, A. G. 1967. On the forms of existence of bacteria in water bodies. Arch. Hydrobiol. 63:238-242.

Shirling, E. B. and D. Gottlieb. 1968. Cooperative description of type cultures of *Streptomyces*. II. Species descriptions from first study. Int. J. Systematic Bacteriol. 18:69-189.

Silvey, J. K. G. 1964. The role of aquatic actinomycetes in self-purification of fresh water streams, p. 227-243. *In* B. A. Southgate (ed.), Advances in water pollution research, Vol. 1, Pergamon Press, London.

——————, J. C. Russell, D. R. Redden and W. C. McCormick. 1959. Actinomycetes and common tastes and odors. J. Amer. Water Works Assoc. 42:1018-1026.

Sloan, P. R. and J. D. H. Strickland. 1966. Heterotrophy of four marine phyto-plankers at low substrate concentrations. J. Phycol. 2:29-32.

Sorkin, Yu. I. 1968. Primary production and microbiological processes in lake Gek-Gel. Microbiology 37:289-296.

Stewart, W. D. P., G. P. Fitzgerald and R. H. Burris. 1967. *In situ* studies on N_2 fixation using the acetylene reduction technique. Proc. Nat. Acad. Sci., U.S. 58:2071-2078.

Stewart, W. D. P., G. P. Fitzgerald and R. H. Burris. 1968. Acetylene reduction by nitrogen-fixing blue-green algae. Archiv. Mikrobiol. 62:336-348.

Straskrabova, V. and M. Legner. 1969. Bacterial and ciliate quantity related to water pollution. Fourth International Conference on Water Pollution Research. Prague. Pergamon Press (Reprint).

Taylor, G. R. and R. K. Guthrie. 1968. Characterization of cytoplasmic antigens for serological grouping of *Streptomyces* species. Bacteriol. Proc. :20.

Thaysen, A. C. 1936. The origin of an earthy or muddy taint in fish. I. The nature and isolation of the taint. Ann. Appl. Biol. 23:99-105.

Trotter, D. M. 1969. A comparison of the carbon dioxide and oxygen rate of change methods for measuring primary productivity. M.S. thesis, North Texas State University.

Vance, B. D. 1965. Composition and succession of cyanophycean water blooms. J. Phycol. 1:81-86.

Venkataraman, G. S. 1960. Growth and pigment changes in *Scytonema tenue* Gardner. Indian J. Plant Physiol. 3:203-211.

Watt, W. D. 1966. Release of dissolved organic material from the cells of phyto-plankton populations. Proc. Roy. Soc., B 164:521-551.

—————————. 1969. Extracellular release of organic matter from two freshwater diatoms. Ann. Bot. 33:427-437.

Wieringa, K. T. 1968. A new method for obtaining bacteria-free cultures of blue-green algae. Antonie van Leevwenhoek J. Microbiol. Serol. 34:54-56.

Williams, A. E. and R. H. Burris. 1952. Nitrogen fixation by blue-green algae and their nitrogenous composition. Amer. J. Bot. 39:340-342.

Williams, S. T. 1966. The role of actinomycetes in Biodeterioration. Int. Biodetn. Bull. 2:125-133.

Wright, R. T. and J. E. Hobbie. 1966. Use of glucose and acetate by bacteria and algae in aquatic ecosystems. Ecology 47:447-464.

Harold L. Allen

Chemo-organotrophy in Epiphytic Bacteria with Reference to Macrophytic Release of Dissolved Organic Matter[1]

Introduction

Lakes of temperate North America are of moderate depth and commonly possess well-defined littoral zones with extensive developments of submerged and emergent aquatic vegetation. Laboratory studies with cultured axenic macrophytes, grown from surface-sterilized seeds under controlled conditions closely simulating the natural environment, have recently demonstrated over brief periods a capacity for extracellular release of over 50% of the photosynthetically fixed carbon as dissolved organic carbon (Wetzel, 1969a, 1969b). Excretory products of macrophytic origin may function as a major source of dissolved organic matter (DOM) in many aquatic ecosystems, especially in low nutrient waters exhibiting poorly developed phytoplankton populations, and where the phytoplanktonic release of extracellular materials is demonstrated to be low. Macrophytic contributions may further be significant in systems where allochthonously derived organic materials are small.

In addition to potential provision of considerable DOM, the aquatic macrophytic vegetation also provides a large surface area for colonization by epiphytic algae and bacteria. Presumably DOM liberated by the aquatic angiosperms would first become available (i.e., as carbon and energy sources) in the immediate nutritional milieu of the attached microorganisms, which in turn may utilize or transform this material prior to its release into the littoral and pelagial regions. Functions and ramifications of the pelagial DOM "pool" and interactions subsequently producing regulatory effects at the primary producer level have been the subject of considerable interest and research during the last several years (see

[1] Supported in part by the U. S. Atomic Energy Commission, Contract AT(11-1)-1599 (COO-1599-21) and the National Science Foundation, Grant GB-6538 to R. G. Wetzel. This paper is a contribution to the Symposium on Structure and Function of Freshwater Microbial Communities, sponsored by the American Microscopical Society, and presented at the American Institute of Biological Sciences annual meeting, Burlington, Vermont, August 21, 1969. Contribution No. 180 of the W. K. Kellogg Biological Station of Michigan State University, Hickory Corners, Michigan 49060. Present address: Department of Biological Sciences, Dartmouth College, Hanover, New Hampshire 03755.

discussion of DOM interactions in relation to lake trophic dynamics and ontogeny of aquatic ecosystems in Wetzel and Allen, 1971).

The purpose of this investigation has been to (1) ascertain the magnitude of direct *in situ* utilization of organic compounds by heterogeneous bacterial communities within the epiphytic muco-organo-polysaccharide complex, and to make comparisons between community responses with respect to depth, time, and macrophyte species (submerged and emergent); (2) to demonstrate under controlled conditions if DOM exosmotically excreted by cultured *Najas flexilis* can be utilized by isolated species of *Caulobacter* and *Pseudomonas* (from natural macrophytes); and (3) to label the emergent *Scirpus acutus in situ* with natural concentrations of evolved CO_2 (^{14}C-labelled) to determine if exogenous carbon dioxide and DOM are released into the epiphytic complex and surrounding littoral area.

The habitat selected for the field portion of this investigation was a small calcareous lake in southwestern lower Michigan (Lawrence Lake, Barry County).

Results

Artificial substrata (1″ x 3″ Plexiglas^R slides) were positioned vertically and horizontally in a random block design in two opposing littoral zones, one of the zones being a marl bench characterized by a stand of *Scirpus acutus*, and the other possessing a highly flocculent organic sediment and characterized by *Najas flexilis* and three species of *Chara*. After suitable acclimatization of substrata (6 weeks colonization), epiphytic samples were routinely removed over a 15-month period (1968-1969) and redistributed into aliquots of ultra-filtered lakewater from the growing site. The maximum rate of bacterial uptake of glucose and acetate-^{14}C was monitored by Michaelis-Menten enzyme kinetic analyses. Respiratory losses of CO_2 during incubation with the radio-substrate were measured on two occasions and were found to contribute an insignificant error to the uptake measurements on an annual basis. A seasonal comparison of chemo-organotrophic uptake parameters from natural and artificial substrata showed no statistically significant differences when rates of utilization were expressed as maximum velocity of substrate uptake (μg 1^{-1} hr^{-1}) per dm^2 of macrophytic surface area at depth. Details of methodology employed are given elsewhere (Allen, 1969a, 1969b).

Rates of uptake of glucose and acetate as μg substrate removed 1^{-1} hr^{-1} dm^{-2} increased towards the sediment in all vertical samples from both sides of the lake. Rates of utilization observed on substrata near the sediments were nearly identical for both substrates on two occasions to rates from similar surface areas of epibenthic (upper 1 mm) samples. Uptake velocities for acetate (vegetative season: 65-130 μg l^{-1} hr^{-1} dm^{-2}; winter season: 8-15 μg l^{-1} hr^{-1} dm^{-2}) were approximately twice as rapid as those observed for glucose throughout the annual period. Similar bimodal (*Scirpus* site) or trimodal (*Najas-Chara* site) peaks in bacterial metabolism were evident for both substrates at all depths. Winter rates of utilization were unusually high in comparison to identical parameters for planktonic bacteria under ice cover in a small Swedish pond (Allen, 1969a), and suggest that the epiphytic bacterial communities are especially well adapted to low temperatures and possess efficient active transport systems for these organic compounds. Further, from the magnitude of annual rates observed, it can be concluded that the epiphytic bacterial chemo-organotrophs are particularly well suited to the direct utilization of low concentrations (0-60 μg l^{-1} added) of both substrates. Such rapid rates of utilization undoubtedly have a pronounced effect upon the littoral DOM "pool," both from the standpoint of quality composition and quantity.

Under aseptic laboratory conditions *Najas flexilis* plants were pulse-labelled with ^{14}C-inorganic carbon for brief periods (< 4 hours) and placed into the center section of a specially constructed 3-part growth chamber, with each of the adjacent sections being separated by pre-eluted, organic-free membrane filters (porosity: $0.22\mu \pm 0.02\mu$). Both species of isolated aquatic bacteria, when placed separately and concomitantly into chamber sections adjoining the labelled macrophytes, were capable of utilizing extracellular products (DOM) of the macrophyte at rates approximating those detected in nature.

Scirpus acutus, when pulse-labelled *in situ* with evolved $^{14}CO_2$ above waterlevel, was found to liberate dissolved organic compounds and exogenous carbon dioxide to the attached microflora beneath the surface. Epiphytes removed with time and depth contained considerable amounts of labelled material; the ^{14}C-label was detectable in the epiphytic complex within a 9-minute incubation period. Samples subjected to micro-autoradiographic analysis showed proportionally higher radioactivity was being incorporated into the epiphytic complex in close proximity to the sediment, a response similar to the increase in chemo-organotrophic uptake with depth. Samples of lakewater analyzed for the liberation of DOM-^{14}C by the macrophyte indicated that up to 65% of the photosynthetically fixed carbon was being extracellularly released over a period of 4 hours in the field (based on triplicate estimates).

The magnitude of extracellular macrophytic release of DOM under natural conditions in the lake, as well as the subsequent rapid incorporation into the epiphytic complex and loss to the surrounding littoral zone suggests (1) the littoral vegetation and attached periphytic growth are capable of significant contributions to the total DOM "pool," and (2) that considerable utilization and transformation of the macrophytically produced DOM is likely to occur prior to its availability in the littoral and pelagial areas.

Literature Cited

Allen, H. L. 1969a. Chemo-organotrophic utilization of dissolved organic compounds by planktic algae and bacteria in a pond. Int. Revue ges. Hydrobiol. 54:1-33.

—————————. 1969b. Primary productivity, chemo-organotrophy, and nutritional interactions of epiphytic algae and bacteria on macrophytes in the littoral of a lake. Ph.D. dissertation, Michigan State University, East Lansing. 186 pp.

Wetzel, R. G. 1969a. Excretion of dissolved organic compounds by aquatic macrophytes. BioScience 19:539-540.

—————————. 1969b. Factors influencing photosynthesis and excretion of dissolved organic matter by aquatic macrophytes in hardwater lakes. Verh. int. Ver. Limnol. 17:72-85.

————————— and H. L. Allen. 1971. Functions and interactions of dissolved organic matter and the littoral zone in lake metabolism and eutrophication. IBP Symposium on productivity problems of freshwaters, Poland. May, 1970.

Robert Benoit, Roger Hatcher, and William Green

Bacteriological Profiles and Some Chemical Characteristics of Two Permanently Frozen Antarctic Lakes

Lakes Bonney and Vanda in the Dry Valley area of Victoria Land, Antarctica (77°32′ S latitude) were cited by Goldman et al. (10) as ideal microcosms to study microbial ecology in oligotrophic systems. Their unusual chemical and thermal characteristics were first investigated by Armitage and House (5). Goldman (9) has recently reviewed the work which has been conducted on Antarctic lakes. Since these land-locked lakes possess unusual chemical and thermal stratification, we hypothesized that the bacteria and fungi in these systems should display a vertical zonation which could be related to their physiological capacities and stress tolerance. Oxygen, ammonia, and sulfide concentrations of these lakes were obtained for the first time. Samples were obtained from late winter (when the September air temperatures 180 cm above the ground varied from $-25°$ to $-46°C$ and the sun was never visible above the horizon) to mid-summer (when the December air temperatures 180 cm above the ground varied from $-7°$ to $+2°C$ and the surface of the ice was continuously exposed to the sun's rays except for several hours in the early morning when they were in the shadow of the Kukri Hills or the Asgard Range.

Procedures

A 10 cm diameter Sipre corer was used to penetrate 4 m ice cover. During the late winter samplings a Scott tent was set up several feet from the hole and samples were immediately processed for chemical analyses. Biological samples were processed minutes after sampling in the hut at Lake Bonney, or at Lake Vanda they were processed in the McMurdo Biolab several hours after acquisition. The sample holes were located at positions that correspond to hole 3 and 23 at Lake Bonney and hole 13 at Lake Vanda of Angino et al. (4). All depths indicated are measured from the bottom of the ice.

Chemical and biological samples were collected in sterile 250 ml vacuated bottles fitted with a rubber stopper and glass-rubber hose connection which terminated with a sealed glass capillary tube. The capillary was suspended 20 cm above the center of the bottle, between two metal supports attached to the sides of the lead-weighted tin container, which held the collecting bottle. The sample container was designed so the bottle was held firmly in place during the descent into the lake, but it could be removed quickly when filled. The sampler was attached to a hydrographic line via an eye hook welded to a horizontal metal brace mounted between the capillary supports 5 cm above the bottle. The capillary was sheared when the messenger dropped down the line. As the hose and shattered capillary sagged away from the sampler, the bottle filled with the uncontaminated sample. Caution was necessary in lowering the sampler into the lake through the ice hole to prevent the line from shearing the capillary.

Oxygen was determined by the azide modification of the Iodometric method (1) as modified for saline waters. Sulfide was determined by the colorimetric method in Standard Methods for the Examination of Water and Wastewater (1). Ammonia was determined by the Nesslerization method (1). Sulfate was determined by the turbidimetric method (1).

Various quantities of water were processed for bacteriological analysis by plating with the spread sample method on media pretempered in petri plates to (0° to 10°C), or cells were collected on 0.45 micron Millipore filters and the filters placed on the media. Duplicate samples were incubated at 15° and 0°C. Several media were used because of the high salt content of the bottom water, which might contain halophilic microorganisms. Media were designated as M-12, M-40, M-41, and HM-1. M-12 contained peptone (Difco) 0.1%, Yeast extract (Difco) 0.1%, 100 ml soil extract and 900 ml ion-free water. Soil extract was prepared by the method of Pramer and Schmidt (14); the soil was taken from a site 100 feet south of the helicopter pad at the Lake Bonney Hut. No organic matter was detected in this soil. This site was lake sediment when the water level of the lake was much higher, hundreds of years ago (Glacial Lake Llano). Lake salts have accumulated on the soil surface as a result of the Dry Valley desert conditions.

M-40 medium contained peptone (Difco) 1.0 g, yeast extract (Difco) 1.0 g, K_2PO_4 0.40 g, $(NH_4)_2HPO_4$ 0.50 g, $MgSO_4 \cdot 7H_2O$ 0.05 g, $MgCl_2$ 0.10 g, $FeCl_3$ 0.01 g, $CaCl_2 \cdot 2H_2O$ 0.10 g, and 1000 ml ion-free water (pH 7.5). M-41 medium contained Trypticase-Soy broth (Difco) 4 g, yeast extract (Difco) 1.0 g, 750 ml of aged sea water and 250 ml of ion-free water (pH 7.0). HM-1 medium contained peptone (Difco) 1.0 g, yeast extract (Difco) 1.0 g, $Na_2HPO_4 \cdot 7H_2O$ 3.35 g, KH_2PO_4 0.89 g, NaCl 1.11g, KCl 0.84 g, $NaSO_4$ 1.78 g, $(NH_4)_2SO_4$ 0.66g, $MgSO_4 \cdot 7H_2O$ 1.23 g, $FeSO_4$ 0.02g and 1000 ml of ion-free water (pH 7.5). All media were autoclaved at 121°C for 15 minutes.

Results

Lake Bonney: Chemistry and Microbiology

Angino et al. (3) have shown that Lake Bonney is a deep (32 m maximum) chemically stratified lake in which sodium and magnesium chloride make up 96 per cent of the dissolved salts. The maximum water temperature of 7°C was observed at the 15 m depth (3) (18). The brine layer at the lake bottom had temperatures which were often several degrees below 0°C in early spring. A warming trend was observed in the 5 to 30 m zone through mid-summer. It is generally agreed that Lake Bonney derives its heat from the sun (15).

Twenty-one per cent of the surface light penetrated the ice cover, and 6 and 2.7 per cent of the light reached the 15 and 20 m depth respectively (10). Therefore, light is not a likely cause of the low biomass, since the ice surface is swept free of snow by the frequent katabatic winds. Approximately 40-50 per cent of the days the senior author spent in the Dry Valleys between August and December, the sky was clear. Goldman observed an average of 30.7 mg $C/m^2/day$ of photosynthetic carbon fixation based upon samples from the 0-10 m depth; in the littoral region a carbon fixation rate of 17.7 mg $C/m^3/day$ was observed (8). There is a very high concentration of oxygen under the ice of both lobes of Lake Bonney (Tables 1 and 2). Supersaturation was observed in the top 15 m of the east lobe in October. The bottom of the water column of the

TABLE 1.—Chemistry of Lake Bonney East Lobe, 1968.

Depth M	Temp.[1] (C)	Specific Gravity[2]	Conductivity[3] (Micromhos)	0^2 (mg/1.)			
				26 Oct.	%sat.	3 Dec.	%sat.
0	0.0	1.0000	—	28.2	201.4	30.5	216.3
5	0.0	1.0005	—	25.6	209.8	24.2	198.4
6	—	—	3,085	—	—	—	—
10	6.0	1.0156	—	12.0	226.4	11.7	220.8
12	—	—	29,410	—	—	—	—
15	7.1	1.1034	140,845	8.1	213.2	1.8	47.4
18	—	—	200,000	—	—	—	—
20	4.9	—	—	2.8	75.7	2.3	62.1
24	—	—	208,335	—	—	—	—
25	1.9	1.1819	—	3.0	81.1	2.8	75.7
28	−0.1	1.1839	212,765	3.1	88.6	3.0	85.7
32	−2.6	1.1983					

[1]Data from ref. 18—Sample 10 January 1965.
[2]Data from ref. 20—Sample 10 January 1965.
[3]Data from ref. 3—Sample 16 December 1961.

TABLE 2.—Chemistry of Lake Bonney West Lobe, 1968.

Depth M	Temp.[1] (C)	Specific Gravity[2]	Conductivity[3] (Micromhos)	O[2] (mg/l.)			
				12 Nov.	%sat.	30 Nov.	% sat.
0	0.1	1.0000	—	24.5	196.0	25.4	203.2
5	0.1	1.0010	—	5.2	74.3	4.0	57.1
6	—	—	2,280	—	—	—	—
10	0.6	1.0545	—	4.3	84.3	2.1	41.2
15	−1.5	—	120,460	2.6	55.3	0.2	0.5
20	−2.8	1.0910	—	0.8	2.2	0.2	0.5
24	—	—	136,985	—	—	—	—
25	−3.0	—	—	0.7	1.9	0.1	0.3

[1]Data from ref. 18—10 January 1965.
[2]Data from ref. 20—10 January 1965
[3]Data from ref. 3—2 October 1961.

west lobe of Lake Bonney was nearly anaerobic during the late November sampling. There was a slight increase in oxygen saturation in the upper layers of both lobes of the lake and a decrease in the oxygen concentration in the lake bottom. The sharpest change occurred at the 15 m depth in the east lobe, where a 166 per cent decrease was observed. No sulfide was detected in the lake but sulfide has been observed in sediment pulled up from the west lobe. It is possible that there is anaerobic activity in the lake bottom when heterotrophic reactions predominate during the austral winter.

Koob (quoted in ref. 9) speculated that there was a dense population of bacteria at 15 m, based upon a high ^{14}C dark fixation at that depth. Goldman et al. (10) observed a dense bacterial band in their direct microscopic counts at 20 m, and they also observed a smaller peak in the 10 m zone. We observed the highest viable bacterial count at the ice-water interface (Table 3). The sea-water medium (M-41) was toxic to most of the bacteria located directly under the ice, as was the HM-1. In all cases tested the M-12 medium was a superior isolation medium for the bacteria located at the ice-water interface, and in the deeper saline areas the HM-1 medium was as suitable as M-12 medium. We have made several attempts to replace soil extract with a defined medium but no combination was as satisfactory as soil extract. Small quantities of vitamins or micronutrients may cause this.

Parts of Taylor Valley may have been under the effects of a marine environment during a previous interglacial period and it is possible that some of the bacteria in the depths of the lake were introduced from the Ross Sea. There was a long lag before bacteria produced observable growth at 15°C during the October sampling. During mid-summer there was a much shorter time lag at the same depth near the same location, and on the identical medium there was an even faster response with bacteria isolated from Virginia lakes. At the end of the Antarctic winter the quantities of available substrate may be very low and as the phytoplankton declines, the bacteria enter the death phase. Those bacteria observed in the early October sampling probably required a long lag to repair depleted metabolic reserves before initiating growth. The lag phase was much more pronounced at 0°C.

The distribution of bacteria in the water column is shown in tables 3 & 4. The bulk of the biomass was located in the 5 m band under the ice in both lobes of the lake. The microbial identification of these cultures will be published elsewhere. These data are not in agreement with the reported dense band of bacteria reported by Goldman et al. (10), but the cells they observed may not have been viable or cultivable on the media used in this study. The high dark carbon fixation uptake observed by Koob may be related to the unusual microflora we observed during the October sampling at the 10 and 15 m depth. At that time most of the colonies observed from those depths were yeasts which belonged to the genera *Candida* and *Cryptococcus*. They were not observed at any other depth. DiMenna (11) observed these genera in Wright Valley soils in the vicinity of Lake Vanda. Benoit (unpublished data) has observed that a high portion of soil microbes in Taylor Valley were yeasts. These yeasts were the first organisms to demonstrate significant growth on the isolation media by several orders of magnitude. However, yeasts were not detected in any of the November or December samplings. The yeasts may be able to utilize the less degradable cell material of the dead phytoplankton during the latter stages of the Antarctic winter. These yeasts

TABLE 3.—Lake Bonney East Lobe (Colonies Per ML).

10/3/68	Incubation Days	M40	M41	HM-1	M12
0 Meters	7	0	0	0	—
0	17	160	0	20	—
10/18/68					
0 Meters	14	250	0	10	308
0	28	300	0	35	354
10	14	15	0	20	29
10	28	18	0	35	33

All plates incubated at 15° C.   Duplicate plates at 0° C. had no growth during 28-day incubation period.

TABLE 4.—Bacterial Counts East Lobe Lake Bonney (Colonies Per ML).

		15°				0°		
10/26/68	Incubation Months	M12	HM-1	M40	Incubation Months	M12	HM-1	M40
0 Meters	1.2	—	40	400	8	—	400	300
5	1.2	—	150	350	8	—	32	400
10	1.2	—	35	35	8	—	20	5
15	1.2	—	25	20	8	—	NGD	NGD
20	1.2	—	20	20	8	—	NGD	NGD
25	1.2	—	10	10	8	—	NGD	NGD
28.5	1.2	—	35	30	8	—	6	6
12/3/68								
0 Meters	4	890	550	—	8	460	370	—
5	4	520	360	—	8	75	115	—
10	4	20	10	—	8	25	42	—
15	4	60	30	—	8	—	30	—
20	4	50	50	—	8	4	43	—
25	4	80	80	—	8	12	3	—
28.5	4	20	20	—	8	3	12	—

may not be able to compete with the bacteria for the organic materials excreted by the algae during the active photosynthetic periods; these yeasts are not as numerically abundant as the bacteria but they may be metabolically more active in the warm 15 m zone (5-8°C). This temperature is 10° to 12°C below the optimum growth temperature of these isolates.

Angino et al. (3) observed that phosphate was at a maximum at 15 m, whereas no phosphate could be detected directly under the ice. We observed no ammonia at any depth. Angino et al. (3) reported that no nitrate could be detected in the lake, but nitrite was detected in the lower regions of the lake with a maximum slightly above 1 mg/1 at 20 m. Goldman (8) observed 0.1 and 0.176 mg/1 of nitrate at 1 and 10 m respectively.

Goldman et al. (10) observed that Lake Bonney contained numerous planktonic forms, less than 20 μ in size, and most of the green coccoid forms with a 10 μ to 5 μ dia were banded at the 15 m depth, while phytoflagellates were clustered at the 10 m depth. *In situ* measurements of bacterial activity and precise measurements of the biomass are needed to resolve the heterotrophic and autotrophic reactions which occur at the thermocline. In the cold saline bottom of the lake the bacterial biomass declined rapidly.

Table 5 records the bacterial viable count of the west lobe of Lake Bonney observed between two November samples. The bacterial count increased in the 0-5 M levels in agreement with the trend in the east portion of the lake, and there was a significant decrease in the samples

TABLE 5.—Bacterial Count (Colonies Per ML) Lake Bonney West Lobe.

12 Nov.*	Depth (M)	Medium M-12	Medium HM-1
	0	180	200
	5	200	200
	10	20	40
	15	35	40
	20	12	20
30 Nov.†	0	250	250
	5	130	280
	10	20	42
	15	15	15
	20	3	6
	24	56	84

*8 weeks incubation @ 15°C.
†6 weeks incubation @ 15°C.

taken from the lower depths. In general the bacterial count in the west lobe was less than that observed in the east lobe. The west lobe is not as deep as the east lobe and it is separated from the east lobe by a narrow, shallow passage. In the west lobe the water is colder and the saline layer nearer the surface of the ice. Most of the meltwater entering the lake probably originates from the Taylor and Rhone glaciers and flows into the west lobe. This flushing action through the narrows may accumulate nitrogen and phosphorus in the east lobe. The ice characteristics in the two lakes are very similar; there is a considerable number of large bubbles throughout the ice which are often 15 cm long and 1 to 1.5 cm in dia at the maximum. In some ice layers there are quantities of algae, especially at the 2 to 2.5 m region below the top of the ice.

Lake Vanda: Chemistry and Microbiology

Lake Vanda (elev. 123 m) is located 25 km north of Lake Bonney (elev. 38 m) and is similar in orientation and topographic setting. Lake Vanda's ice cover has a thickness and surface features similar to those of Lake Bonney, but the water column beneath the ice is very different. In Lake Vanda the ice bubbles observed in Lake Bonney were not observed. Wilson's model of Lake Vanda divides it into three zones; in the top 10 m the temperature is lower than 5°C and weakly buffered (19). This lake is one of the clearest known, with an extinction coefficient as low as 0.031 for blue light (10). In the zone between 20 and 40 m there is a saline circulating zone with a high buffer and a temperature which remains close to 8°C throughout (19). In the 40 to 65 m zone the buffer capacity increases dramatically and the pH plateaus at 6.6; this region is non-convective for heat flow considerations and a Ca-Mg chloride brine predominates. A maximum temperature of 25.1°C has been observed at the bottom. Wilson and Wellman (19) considered Lake Vanda to be a solar heat trap. Ragotzkie and Likens (15) concluded that geothermal heating was largely responsible for the high brine temperatures, with solar heat providing additional heat during the summer. The oxygen levels are reported in table 6. Supersaturation was observed from the 0 to 50 m level, but the degree of supersaturation was less than the water directly under the ice at Lake Bonney. At 55 m we observed a sharp increase in salinity and sharp drop in oxygen concentration at the deepest point. The bottom 5-10 m of Lake Vanda is probably anaerobic much of the year. The ammonia and sulfide data are shown in table 7; the values reported therein are typical end products of anaerobic microorganisms. The ammonia may be produced as a result of ammonification of dead cell material. The sulfide may be the result of sulfate-reducing bacteria or geochemical sources. Sulfide odor was reported from samples below 61 m by Angino et al. (4) but was not measured. Yamagata et al. (20) did not report any sulfide in Lake Vanda. In table 8 the variations of the sulfate concentrations are indicated. The Na, Cl, and Ca horizontal profiles are relatively constant during November, December, and January but the sulfate values show wide variation (4). During November, December, and January, sulfate concentrations were 300, 7,000, and 2,000 mg/1 at the 60 m depth. These values are higher than those reported here, but this may be due to differences in the state of the system at different

TABLE 6.—Some Chemical Characteristics of Lake Vanda.

Depth (M)	Temp.[1] (C)	Conductivity[2] (Micromhos)	Ca[2]	O_2 (mg/1.) Avg. of 5 Measurements from Nov. to Dec.	
					% Saturation
0	0.1	—	—	17.1	120
9	—	585	46	—	—
10	5.0	—	—	17.4	141
20	8.0	—	—	16.6	136
21	—	1,990	182	—	—
30	8.0	2,025	182	16.4	137
39	—	2,025	182	—	—
40	8.6	—	—	15.8	134
50	14.8	—	—	16.1	196
51	—	22,220	2,730	—	—
55	20.5	—	—	0.3	4.8
57	—	71,430	11,700	—	—
60	24.2	100,000	16,700	0	0

NOTES:
[1]Data from ref. 18—4 January 1965.
[2]Data from ref. 4—1 January 1965.

TABLE 7.—Ammonia and Sulfide Concentration, Lake Vanda, 1968.

Depth M	NH_4 (mg/1.)[1]	S (mg/1.)[2]
0	0	0
10	0	0
20	0	0
30	0	0
40	1	0
50	4.9	0
55	8.2	1.3
60	12.8	22.0
65	—	28.4

NOTES:
[1]Sample of 7 December 1968.
[2]Sample of 14 December 1968.

TABLE 8.—Sulfate Concentrations of Lake Vanda, 1968.

Depth M	13 Nov.	7 Dec.	14 Dec.
0	12	15	15
10	17	20	22
20	19	21	27
30	36	47	48
40	48	50	50
50	75	406	500
55	380	475	660
60	410	500	780

sampling times, or differences of sampling site away from the deepest point. Values indicated in table 8 are in agreement with those of Yamagata et al. (20). The variation in the sulfate content may be related to not only geochemical conditions but changes in the annual biological cycles relative to the sulfur cycle.

The bacteriological data are shown in table 9. The hole sampled in the October 9 sampling was about 1/8 mile southwest of the site used on December 7 and the depth at this point was only 40 m. In October few bacteria were present under the ice surface in contrast to the situation at Lake Bonney, where the count was approximately 400/ ml. The bacteriological content increased with depth and the differences were greatest on the M-12 medium. In the December sampling there were two peaks of bacteriological activity, i.e. directly under the ice and at the 40-50 m depth. These two zones correspond to the zones of maximum primary productivity reported by Goldman (8) (10). The transient nature of the bacterial population under the ice may correspond to a fresh water algal bloom such as Bunt has reported under sea ice at nearby

TABLE 9.—Bacterial Counts, Lake Vanda (Colonies Per ML).

		15°			0°	
10/9/68	Incubation Months	M12	HM-1	Incubation Months	M12	HM-1
0 Meters	0.3	0	0	7	2	6
10	0.3	20	0	7	40	0
20	0.3	30	10	7	15	12
30	0.3	200	15	7	45	32
40 (Bottom)	0.3	150	10	7	175	65
12/7/68						
0 Meters	4	400	20	6	310	20
10	4	50	10	6	25	10
20	4	25	10	6	12	12
30	4	20	15	6	5	3
40	4	200	60	6	180	35
50	4	200	20	6	175	10
57 (Bottom)	4	0	0	6	0	0

McMurdo Sound (7). No bacteria were isolated at the 57 m depth but this may have been due to the anaerobic nature of the region. No attempt was made to isolate anaerobic bacteria. Sugiyama et al. (16) observed 117 strains of fungi from Lake Vanda. We did not observe any yeasts. The M-12 medium is suitable for yeast primary isolation only when they are numerically the dominant form; otherwise the bacteria overgrow them. We did not observe any of the filamentous molds reported by Sugiyama et al. (16). The M-12 medium will support the growth of all *Penicillium* and *Aspergillus* spp. indicated. Sugiyama et al. (16) did not indicate the density of the fungi and the isolation was made one week after samples were returned to Tokyo; therefore, the fungi are probably present in low numbers.

Discussion

Lakes Bonney and Vanda are marked not only by sharp geochemical differences, but by pronounced microbial ecological differences. A meromictic condition was clearly demonstrated when Armitage and House (5) pumped 100 gallons of lake water through a 25 mesh plankton net from each lake, and isolated only one *Chlorella*-like alga. Meyer et al. (13) isolated no microorganisms from the 60 m depth of Lake Vanda but they used aerobic culture methods, as used in this work. Anaerobic culture methods must be used in future investigations. A chemotactic mechanism might permit strict anaerobes to retain their position with the ideal substrate and sulfide concentrations. Little is known about phototactic mechanism *in situ* but the observation of a bacterial maxima in the deep part of a lake over the thermocline or the lower layers of the epilimnion has been shown by Stangenberg (17) in four Italian lakes. Goldman et al. (10) observed that the maximum primary productivity on February 14, 1963 was observed in the 60 m zone. The anaerobic chemosynthetic bacteria could account for this phenomenon. They were unable to establish any species trends in zone because of the low plankton counts. The bacteriological bloom observed under the ice in December chronologically parallels the peak of light fixation observed by Goldman (8) December 28, 1961.

At Lake Bonney the bacterial biomass maximum was located directly under the surface of the ice; the magnitude (and apparently the activity) of this population increases as the peak of the austral summer is reached. The littoral waters of Lake Bonney have a greater quantity of life than those of Lake Vanda (8). Goldman (8) observed that nitrate doubled the primary productivity of the littoral waters of Lake Vanda, whereas phosphorous had little effect. These data, in conjunction with negative ammonia and nitrate values and slight nitrite values, indicate nitrogen is the limiting element in the lakes. Holm-Hansen (11) has demonstrated that *Nostoc commune* was present in the soils of meltwater streams near the lake, and that it can fix nitrogen under laboratory conditions. The lake *in situ* nitrogen fixation rates are unknown. Benitrification reactions in the anaerobic portion of the lakes could prevent any accumulation of nitrogen in the system. Lake Bonney may be a more productive lake because its large deep east lobe is not anaerobic to the degree Lake Vanda is. The lake sediments may be an important cell in the nitrogen pool of these lakes. There is need of frequent and detailed studies of the nitrogen and phosphorus transformations of the different zones of the lake.

The sulfur cycle may play a dynamic role in the annual cycling of carbon and other nutrients in Lake Vanda. During the winter when heterotrophic reactions predominate, the dead cells of the phytoplankton and zooplankton would accumulate at the chemocline or thermocline of the lake. Under anaerobic conditions the fatty acid end-products of the anaerobes would provide the energy and carbon source for the sulfate-reducing bacteria to produce H_2S. When light is available during midsummer the photosynethic bacteria will couple the energy-yielding light reactions with the dark fixation reactions, and hence produce more cell material, which will be available for decomposition during the winter. The purple and green sulphur bacteria require reduced sulfur compounds as electron donors, and if they are active the H_2S concentration should decrease during the austral summer and the sulfate concentration should increase. The necessary shifts in sulfate concentration have been observed by the authors and Angino et al. (4). Unfortunately, we have only one sulfide reading. Benoit and Hall (6) have isolated a number of sulfate-reducing bacteria from freshwater pools of Taylor, Victoria and Miers Valley; therefore, they are ubiquitous in the valleys.

There are many geochemical questions about Lake Vanda which remain to be answered, and the variations may be explicable on a purely nonbiological basis. However, the biological cycling of carbon and sulfur at the surface of the nonconvective warm brine is an equally attractive theory. Less than one per cent of the incident light is available for photosynthetic reactions at the 55 m level, (10) so the light reactions must be efficient during the brief austral summer if the winter-summer cycling of sulfate and sulfide is stoichiometrically linked to the carbon fixation and heterotrophic reactions of this unusual oligotrophic system.

Acknowledgments

The authors express their appreciation to the Office of Antarctic Programs, National Science Foundation, for grant GA-323230-0, which made this study possible, and to the officers and men of the Antarctic Support Force and VX-6, for their logistical support.

Literature Cited

1. American Public Health Association, (1965) Standard Methods for the Examination of Water and Wastewater, 12th Edition, APHA, New York, N.Y., 626 p.

2. Angino, E. E. and K. B. Armitage. (1963). A Geochemical Study of Lakes Bonney and Vanda, Victoria Land Antarctica, J. of Geology 71: 89-95.

3. Angino, E. E., K. B. Armitage and J. C. Tash. (1964) Physiochemical Limnology of Lake Bonney, Antarctica. Limnol. Oceanogr. 9: 207-217.

4. Angino, E. E., K. B. Armitage and J. C. Tash. (1965) A Chemical and Limnological Study of Lake Vanda, Victoria Land Antarctica, Kans. Univ. Sci. Bull., No. 45, 1097-1118.

5. Armitage, K. B. and H. B. House. (1962) A Limnological reconnaissance in the area of McMurdo Sound, Antarctica. Limnol. Oceanogr. 7: 36-41.

6. Benoit, R. E. and C. L. Hall, Jr. (1970) The Microbiology of Some Dry Valley Soils of Victoria Land Antarctica. *In* "Antarctic Ecology," Ed. Holdgate, M. W., Academic Press, New York p. 697-701.

7. Bunt, J. S. (1964) Primary Productivity under Sea Ice I. Concentration and Photosynthetic Activities in Microalgae in the Waters of McMurdo Sound, Antarctica. *In* Biology of the Antarctic Seas, Ed. by M. O. Lee, Vol. I, Antarctic Research Series, American Geophysical Union.

8. Goldman, C. R. (1964) Primary Productivity Studies in Antarctic Lakes. *In* "Biologie Antarctique: Antarctic Biology". R. Carrick, M. W. Holdgate and J. Prevost. eds. Hermann, Paris. p. 291-299.

9. Goldman, C. R. (1970) Antarctic Freshwater Ecosystems. *In* "Antarctic Ecology" (Holdgate, M. W. ed.) Academic Press, New York. p. 609-627.

10. Goldman, C. R., D. T. Mason and J. E. Hobbie. (1967) Two Antarctic Desert Lakes. Limnol. Oceanogr., 12:295-310.

11. Holm-Hansen, O. (1964) Isolation and Culture of Terrestrial and Freshwater Algae of Antarctica. Phycologia 4: 43-51.

12. Menna, M. E. di (1960) Yeasts from Antarctica. J. gen. Microbiol. 23: 295-300.

13. Meyer, G. H., M. B. Morrow, O. Wyss, T. E. Berg and J. L. Littlepage (1962) Antarctica: The Microbiology of an Unfrozen Saline Pond. Science 138: 1103-1104.

14. Pramer, D. and E. L. Schmidt. (1964) Experimental Soil Microbiology. Burgess, Minneapolis, Minn.

15. Ragotzkie, R. A. and G. E. Likens (1964) The Heat Balance of Two Antarctic Lakes. Limnol. Oceanogr. 9: 412-425.

16. Sugiyama, J., Y. Sugiyama, H. Iizuka and T. Torii (1967) Report of the Japanese Summer Parties in Dry Valleys, Victoria Land, 1963-1965. IV. Mycological Studies of the Antarctic Fungi Part 2 Mycoflora of Lake Vanda, an Ice Free Lake. Antarctic Record, No. 28 p. 2247-2256.

17. Stangenberg, M. (1966) Epi-Metalimnetic Maxima in Vertical Distribution of Bacteria in Some Subalpine Italian Lakes. Mem. 1st Ital. Idrobiol. 20: 105-115.

18. Torii, T., N. Yamagata and T. Cho. (1967) Report of the Japanese Summer Parties in Dry Valleys. Victoria Land, 1963-1965. II. General Description and Water Temperature Data for the Lakes. Antarctic Record, No. 28, p. 2225-2238.

19. Wilson, A. T. and H. W. Wellman. (1962) Lake Vanda: An Antarctic Lake. Nature Lond. 196: 1171-1173.

20. Yamagata, N. T. Torii and S. Murata. (1967) Report of the Japanese Summer Parties in Dry Valleys. Victoria Land, 1963-1965. V. The Chemical Composition of Lake Waters. Antarctic Record, No. 29, p. 2339-3361.

David G. Frey

Concluding Remarks

This symposium was sponsored by the American Microscopical Society, which has as its stated purpose the encouragement of research involving the microscope, microscope accessories, and microscopical techniques. Although all the papers were concerned with freshwater microorganisms, in relatively few of them were a microscope or microscope accessories or even microscopical techniques used as major items of hardware or procedure in data collection. A show of hands revealed that considerably less than a majority of the audience were members of the Society, and at concurrent business luncheons of the AMS and the Freshwater Section of the Ecological Society of America, held in adjoining rooms, persons at the AMS meeting constituted a virtually non-overlapping population with the symposium audience, most of the latter being at the neighboring meeting.

Organization of the program at the annual meeting of the Microscopical Society is the responsibility of the Vice-President, in this case Dr. John Cairns, who, although he personally uses a microscope in his work, is certainly much more of an ecologist than a microscopist. The symposium reflects his interests. It could have been sponsored equally well, and perhaps more appropriately, by the Ecological Society or the American Society of Limnology and Oceanography. Quite obviously we need to encourage more joint meetings of these several societies to coordinate overlapping interests and to keep a significant symposium such as this from becoming buried in some obscure place where interested persons might not think to look for it.

The present symposium was scarcely buried. It attracted a large audience, most of whom were faithful to the end, which attests to the superior quality and high interest level of the papers. Having them together in one volume where they can be studied at leisure should help to stimulate additional research on freshwater microorganisms and their functions in aquatic ecosystems. Certainly commendations should go to Dr. Cairns for organizing this fine symposium and to all the participants individually for jobs well done.

To most limnologists, the term—microorganisms—calls to mind such names as *Daphnia, Diaptomus, Keratella, Ceratium, Asterionella*—organisms that are large enough to be collected with a plankton net and to be studied conveniently with an ordinary light microscope, and diversified enough morphologically to be assigned with reasonable certainty to morphospecies. Yet these organisms were scarcely considered at all in the symposium, except as members of broad functional groups. Instead, major attention was directed toward the bacteria, nanno-algae, and protozoans, and to a lesser extent, the fungi.

Significantly, nearly all the papers presented were process-oriented rather than organism-oriented. Microbial systems were used to study broad ecological problems, both in the field and laboratory. Cooke argued for the use of laboratory microcosms to study basic principles of ecology at different levels of organization, because such systems can be controlled, manipulated, and replicated with ease to yield repeatable results over short time intervals. Successional changes in such ecosystem properties as production, biomass, and structural and metabolic diversity are amenable to study in laboratory communities, and the results are directly relatable to the large and more complex ecosystems in nature. In fact, Salt in his discussion of strategies for the elaboration of ecological theory argued that if a response or interaction is general, then it should be observable in Protozoa and other simple animals. Further, if a particular phenomenon has been observed in both Protozoa and higher animals, its manifestation in the Protozoa is probably simpler and more readily comprehensible. Taub used a laboratory community consisting of 1 alga, 2 bacteria, and 1 protozoan in an attempt to answer some general questions.

Odum proposed the interesting hypothesis, and quoted a variety of empirical observations in its support, that light might be used in respiration as well as in, or as an alternate to, photosynthesis. In stressed situations where autochthonous organic matter tends to accumulate or where there is a high input of allochthonous organic matter, as in pollution, a metabolic blockage develops through the accumulation of nutrients in bound organic form. Normal respiration is not sufficient to recycle the organic matter rapidly. Under such conditions the chlorophyll content of the plants decreases relative to other pigments, such as carotenoids, which possibly facilitate absorption of light energy in the accomplishment of photorespiration or photoregeneration. Odum suggested that the chlorophyll/carotenoid ratio indicates successional stages, as Margalef proposed, only under rather special conditions, and that more usually this ratio reflects the balance between photosynthesis and photorespiration; it might be considered to represent the extent of switching of light energy to one half system or the other, depending on whether photosynthesis or respiration is limiting. One of the potential consequences of this mechanism, if it is as important as Odum predicts, is that in systems where both photosynthesis and photorespiration are high, estimates of photosynthesis by procedures commonly used today may yield almost meaningless results. Because this paper came early in the symposium, other participants attempted to relate their own studies to it. Benoit, for example, pointed out the high percentage of chromogens among the bacteria in two permanently frozen lakes in the Antarctic, and a number of speakers (Cooke, for example) alluded to the decline in the chlorophyll/ carotenoid ratio that occurs as unialgal cultures proceed to senescence or as complex communities proceed to maturity.

Saunders is attempting a systems analysis of the carbon flow in aquatic systems by measuring the carbon content of the various functional compartments at appropriate short intervals of time, such as sunrise and sunset, and the transfer rates between compartments. He has identified a number of coupled subsystems, each with its own controls, oscillations, and lags, for all of which the 24-hour day-night cycle acts as a forcing variable. For these reasons a steady-state condition is perhaps never realized, and likewise there are seldom any simple correlations between fluctuations in various components of the system. The three coupled subsystems identified to date are phytoplankton-dissolved organic matter-bacteria, phytoplankton-herbivorous zooplankton, and diel changes in the protein-carbohydrate composition of algae, the latter subsystem operating at the molecular level. Microorganisms control the movement of carbon in the system, even though inorganic carbon and dissolved organic matter (DOM) are the biggest compartments. Moreover, the microorganisms have some capacity to regulate their change of state, and hence are not limited to passive response to environmental variables.

A significant fraction of the carbon fixed in photosynthesis, both by algae (Fogg and his associates) and macrophytes (Wetzel, Allen, and associates), is released directly into the environment as dissolved organic matter. Studies on the uptake of specific organic compounds from the DOM pool by tracer techniques developed by Hobbie, Wright, and their students indicate that probably a large percentage of the primary productivity and of allochthonous organic matter passes through bacteria, that the uptake velocity varies directly with the concentration of organic matter present in the DOM pool, being several orders of magnitude lower in oligotrophic than in polluted situations, for which reason the maximum uptake velocity may prove to be a convenient means of placing a lake in the trophic continuum, and that removal of organics from the pool by bacteria is so efficient that concentrations seldom become great enough to enable algae to live heterotrophically. Uptake of specific organic substances depends on specific transport systems, which can be inactive between peaks of organic matter (e.g., plankton blooms) and can rapidly be reactivated with the reappearance of the proper substrate. The formation of particulate matter by bacteria can be greater than by algae, and bacteria can represent the first particulate form of some of the carbon fixed in photosynthesis through uptake from the DOM pool. The latter is potentially very significant in terms of ecosystem energetics and food-chain dynamics.

Woodring was concerned with two other coupled subsystems mentioned by Saunders. Laboratory cultures of algae produced carbohydrates during light periods and protein during the dark. The diel migration of zooplankton is thought to be adaptive by enabling the animals to feed at a higher temperature (near the surface) and subsequently to digest the food, grow, and reproduce at a lower temperature. An additional or alternative adaptation is that by feeding at night in the trophogenic zone, the zooplankton obtain a higher percentage of protein in their food. Reciprocally, the algae are able to satisfy at least a part of their nitrogen requirements with ammonia (from the zooplankton), which would entail a smaller energy cost in synthesizing organic matter than with nitrate as the source.

Some of the significant advances in ecology today are coming from the study of the invasion and colonization of isolated habitats—islands. These can be islands in the usual sense surrounded by water and located at varying distances from the mainland, completely new land masses such as Surtsey off Iceland, glass slides, polyurethane sponges or other artificial substrates placed in lakes or streams, or bottles of water placed at progressively greater distances from a suspected source of dissemules of aquatic organisms. Both diatoms and protozoans in fresh water exhibit a high species diversity but low equitability. The number of colonizing species bears a direct relationship to the area (diatoms) or volume (protozoans) provided, and species-frequency plots in each group yield a truncated log-normal distribution, with the mode occurring among species having low numbers of individuals in the total population. Community responses to environmental conditions is assessed by the change in shape of the curve. For diatoms (Patrick) the effect of both pollution and reduced current velocity, although for different reasons, is to reduce the height of the mode, meaning that the proportion of species represented by only a few individuals in the population is decreased. The factors responsible for maintaining a high species diversity in diatoms are the size of the area available for colonization, the size of the species pool, and the reinvasion rate.

In the case of freshwater protozoans there is also a large number of species in the pool, most of which are widely distributed geographically. Yet comparable habitats unaffected by serious perturbations tend to have about the same number of species, although the particular species present seem to vary at random except for a few of the most common species.

Cairns has developed a model with a finite number of niches, each of which can be occupied by any one of a number of species having similar requirements. The communities are conceived of as having an internal structure with appropriate interactions between species, but with the species composition of the community largely controlled by external factors or even by purely fortuitous circumstances of colonization. The dynamics of invasion, colonization, and extinction can be studied readily in such micro islands.

Maguire has been concerned with the successional development and increasing complexity of whole communities of microorganisms, not with the dynamics of particular taxocenes as are Patrick and Cairns. His islands —aquaria on Surtsey, jars of water, or leaf axils of bromeliads—represent isolated aquatic microhabitats, in which colonization can occur only by aerial dispersal of dissemules or by transport of them by larger animals, such as raccoons and insects, that may have access to the microsystems. Maguire feels that such systems better represent real-world conditions than do microcosms in the laboratory.

A most fascinating suggestion from the symposium is that bacteria might be able to synthesize B vitamins in clouds. Parker found in rainfall significant amounts of cobalamine, biotin, and niacin, which, especially cobalamine, could be adequate to support the development of blooms of algae in vitamin-deficient situations. He postulated that nascent rain drops about 100 μ in diameter could serve as miniature aquaria, enabling the bacteria to survive in clouds long enough to produce some vitamins. Freezing of the water droplets just prior to precipitation could result in the release of vitamins from the cells. The source of the vitamins in rainfall is still speculative, however. Biotin and niacin are possibly associated with pollen rains, but not cobalamine.

Four papers concerned with seasonal cycles and functional relationships in various natural situations make one realize how little we know about some of the major groups of aquatic microorganisms that are obviously very important in limnological processes. Even their names are unfamiliar.

The chance finding of *Thiopedia* in a small pond near the Kellogg Biological Station led Hirsch into a detailed study of the microorganisms in the pond. Imaginatively using some old techniques, such as glass slides and plastic films, and some new techniques, such as formvar-coated EM grids and direct observation with water-immersion objectives, Hirsch identified a total of 241 kinds of microorganisms, 82 of them bacteria, distributed among nine communities in the pond, and attempted to get some understanding of their seasonal dynamics. Yeasts and iron bacteria, for example, were abundant beneath the ice in winter.

Benoit reported on his studies of microorganisms in two permanently frozen lakes of the Antarctic, which differ considerably in their chemical and thermal characteristics. In Lake Vanda large populations of bacteria occur at the bottom throughout the year but toward the surface only in summer. Benoit postulated that the upper population depends directly on algae, which are seasonal, whereas the bottom population is sustained by the long-term accumulation of sedimented organic matter. A large population of yeast occurred beneath the ice of Lake Bonney over winter.

Silvey and his associates have been studying the seasonal cycles of the algae, bacteria, and actinomycetes in two reservoirs of the Southwest for a number of years, with some attempts at identification, especially of *Streptomyces*, and some attempts to work out functional relationships. The seasonal cycles are repeated in general pattern from one year to another. The gram-negative bacteria and the bluegreen algae peak about the same time in August, followed by the actinomycetes, and then by gram-positive bacteria. The almost reciprocal relationship between the latter two groups suggests that the gram-positives either attack the actinomycetes directly or at least metabolize organics produced by them. Because of their abundance and seasonal cycles in relation to other groups of microorganisms, Silvey believes the actinomycetes must be very important in freshwater dynamics.

Even less is known about the occurrence and roles of the true fungi (mainly phycomycetes) in freshwater systems. Paterson, in a thorough review of the literature as well as of his own current research, reported that chytrids sometimes actively parasitize growing algal cells, and thereby may help regulate population size, whereas in other instances, they attack only senescent or already dead cells. In some instances infested populations or portions of populations formed distinct horizons in the water that moved downward with time. Such biological "tagging" might be useful in studying the sinking rates of various species of algae in relation to turbulence and other factors. Chytridiaceous and oomycetaceous fungi are abundant in sediments and, since at least some of them exhibit a preference for culture substrates containing chitin, very likely they are important in the digestion of chitin.

Finally, McKinney pointed out that our present methods for the microbiological treatment of wastes—trickling filters, activated sludge, oxidation ponds, and anaerobic digestion—were designed chiefly by engineers having little or no real comprehension of the biology involved. Urgently needed is the direct and immediate attention of biologists in studying these processes to make the present systems more efficient, and hopefully to develop even better ones, if the challenge of environmental pollution is to be met.

EDITOR'S NOTE

The manuscripts for the following papers presented at the symposium were unfortunately not available at the press deadline date of August, 1970, approximately one year after the symposium date:

"Structure and Function of Bacterial Communities in a Forest Pond, the Habitat of *Thiopedia* sp." by Peter Hirsch, Michigan State University, East Lansing, Michigan.

"Control in Aquatic Ecosystems," by Douglas Woodring, University of Michigan, Ann Arbor, Michigan.

It is regrettable that our printing schedule did not permit further delay so that these fine papers could be included. This note is included to explain the absence of certain papers presented at the symposium. It should not be construed as criticism of the authors of the missing manuscripts who were beset by personal problems, such as illness, which made completion of the manuscripts impossible.